YOUR DOG

CASSELL'S PET AND FANCY BOOKS

1. HAMSTERS by Percy Parslow
2. PONIES by Veronica Heath
3. RACING PIGEONS by Guy Barrett
4. FANCY MICE by R. S. Hutchings
5. BUDGERIGARS by W. Watmough
6. CAVIES by Alan Sole
7. BANTAMS by H. Easom Smith
8. GOATS by H. E. Jeffery
9. DONKEYS by Robin Borwick
10. CATS by Grace Pond and Elizabeth Towe
11. FANCY PIGEONS by W. Watmough
12. BEES AND BEEKEEPING by A. V. Pavord
13. RABBITS by H. Dyson
14. SIAMESE CATS by Elizabeth Towe
15. TROPICAL FISH AND AQUARIA by Leslie B. Katterns
16. CANARIES by George Lynch
17. EXOTIC BIRDS FOR CAGE AND AVIARY by Clive Roots
18. GARDEN PONDS AND INDOOR WATER GARDENS by Leslie B. Katterns
19. YOUR DOG by Frank Jackson and W. Ronald Irving

YOUR DOG

by Frank Jackson
and W. Ronald Irving

with a Foreword by Joe Cartledge
International Judge
and Canine Journalist

CASSELL · LONDON

CASSELL & COMPANY LTD
35 Red Lion Square, London WC1R 4SJ
Sydney, Auckland
Toronto, Johannesburg

First published 1972

I.S.B.N. 0 304 93810 6

Typeset by
H. Charlesworth & Co. Ltd.
Huddersfield
Printed by
Redwood Press Ltd.
Trowbridge, Wiltshire, England
F. 1171

CONTENTS

FOREWORD BY JOE CARTLEDGE

Apart from books on sex I would think it true to say that more books are written these days on dogs than on any other one subject. Many are concerned with a particular breed of dog, and others relate to different aspects of the canine world. This, however, is a book with a difference, a mine of information, written in a style which makes it good reading, and no dog owner or would-be dog owner can help but benefit by sitting down and taking it in. This book makes me, as a dog journalist of sorts, wish that it was my brain child and that I had come up with the idea first of having such thoughts published in book form. If by good chance this book is read before the reader and his family purchase a dog for themselves, so much the better. As the authors so rightly say, you can't choose a wife or husband for anyone, and neither can you choose a pet. But the concise descriptions of the breeds from which you have freedom of choice will give you a very good idea as to the breed of pet you and your family are most suited to own. A person starting to breed dogs can, I feel, also benefit much from this book by two writers who have been on the dog scene, with pet dogs, show dogs, as well as hunting with dogs, from boyhood. So many books in the past and over the years have been written by people who were writers first and dog men second. In this case it is not so. Beginners often get so many contradictory and conflicting opinions on every aspect of dog ownership that they are left completely confused and in a state of bewilderment. This is a practical book for practical people. A book that will neither confuse nor mesmerize, but a book that will make every reader more suitable to own a pet dog or to start out as a breeder. A happy healthy dog is what every owner or would-be owner is looking for — this book can be your bible.

PREFACE AND ACKNOWLEDGEMENTS

This book is, we hope, different from the very many others which deal with the subject of keeping a dog. We have tried to write a book which emphasizes the responsibilities involved in keeping a dog, whereas others seem to be principally interested in selling dogs and in making an already popular pet even more popular. We hope that we have not written yet another P.R.O. blurb for dogs; we don't believe that they need it and in any case our interest in dogs is not on this superficial level. We hope that we have written a book that will help people to choose a dog wisely, to give him the attention that he deserves and to derive from him the enjoyment and interest that we derive from our dogs.

During the course of writing this book we have received help from many people, who have given freely of their time and advice. We are particularly indebted to Jean Jackson, whose knowledge of a dog's nutritional requirements and whose practical experience of rearing puppies has been of immense help to us. While we were writing the book she gave us the benefit of her advice and knowledge, and when it was written was able to decipher our manuscript and in the course of typing it was able to correct our more outrageous statements.

We should also like to thank Anne Roslin-Williams, for the photographs on pages 2, 10, 13, 20, 25, 29, 50, 51, 76, 79, 91 and 96, and Anne Cumbers for those on the front cover (showing a family of Sussex Spaniels, owned by Mrs. J. Munday) and pages 3, 5, 8, 17, 44, 73, 77, 81 and 84; the drawings and the frontispiece were supplied by Frank Jackson.

Frank Jackson
Pease Pottage
April 1971

W. Ronald Irving
Musselburgh

CHAPTER 1
Choosing a Breed

The relationship which exists between dog and man is a unique one which has developed over many thousands of years but which reflects appreciably more credit on the dog than it does on man. The bond stemmed from a mutual appreciation by prehistoric man and the wild dogs which were his contemporaries, that their varied hunting skills were complementary and could be used to the advantage of both partners in the association. As the association developed man assumed the role of dominant partner and as his society progressed he began to demand more of the dog than the benefit of his inherent hunting skills. Dogs began to be used for guarding the home or the herding of domestic cattle and for a variety of other utilitarian purposes. It was not until civilization had progressed to such an extent that man had the opportunity to indulge in activities which were not directly concerned with the task of keeping body and soul together that dogs began to appear and be retained as part of the household which could not undertake any of the tasks that were traditionally allotted to them. Dogs began to be appreciated as companions and as symbols of wealth—and even assumed the role of gods.

As a consequence of the loss of contact with functional requirements a great diversity of shape, colour, size and general appearance emerged, and dogs which exhibited some novel feature or whose normal physical characteristics were in some way distorted were prized for the attention that ownership of such an oddity could bestow on their owners. This trend has continued to the present day but has not been and is not always in the best interest of the dogs themselves.

Once contact is lost with the job of work for which a particular breed was developed and which demanded that that breed must be absolutely fit and strong, should be

An English Springer Spaniel, owned by Mrs Sherwood and Miss Damerell, retrieving over high wire. Continued contact with the purpose for which a breed was originated is a valuable safeguard of soundness and integrity.

sagacious in pursuit of its allotted task, and amicable towards others with whom it was obliged to work, then the way was, and is, open for those who take up the breed, but fail to understand it, to affect changes in the name of 'improvement' which will eventually usurp all the breed's original characteristics. However although this sad fate has overtaken many breeds it has not affected them all, and although the opportunity to use the breed for the purpose for which it was developed is now denied to many enthusiasts they jealously guard the breed's original characteristics and protect its physical soundness.

In an age when many people live totally urban lives ownership of a dog may represent the only opportunity for close contact with another animal species, or for the lonely may be the only close contact with any living thing. However not all dogs are acquired simply to provide companionship; the reasons for getting a dog are many and varied. Pressures are brought to bear on people to get a dog which in some subtle way will enhance their social status, and the choice of dog is often made without any consideration being given to the needs of the dog itself. People who are out at work all day buy a dog, only to be horrified at the damage which a bored and confined dog can do to the home. People without the facility to exercise a dog may acquire a breed which was originally bred to hunt over many miles of open country. The nervous or hesitant may acquire a guard dog which they are subsequently unable to control, or people with young children may be attracted to a breed which does not readily tolerate children. Before you acquire a dog you must be quite sure that you are prepared to accept all the responsibilities that ownership imposes and that the dog that you select is of a breed which is suited to the life that it is to lead.

However, choosing a particular breed is rather like choosing a wife; it cannot be successfully done for you by some third party. Although you must be aware of all the factors which affect your choice, it cannot be an entirely logical process. It is quite possible that an attraction to a particular breed can be sufficiently great for the owner to change his habits to suit the requirements of the selected

A good pack of hounds like these Beagles subdues individuality in order that the pack can work as a closely integrated unit. They are lively, sociable and totally dedicated to their one purpose in life.

breed. The habitually sedentary person may become active by becoming the owner of a dog which requires considerable exercise, and one has only to notice the number of ill-groomed women, and men, at dog shows lavishing their attention on a superbly groomed dog to appreciate that sacrifices are often made for a particularly captivating breed.

The only way that a breed can be chosen is to see as many dogs as possible, but before we examine how this might conveniently be achieved we ought to discuss the thorny question of whether it is better to get a pedigree or a mongrel dog. It cannot be disputed that if a dog is expected to do a job of work then it should be of a breed developed to do that work and from parents of proven ability. Many dogs however are acquired simply as pets, and choice is considerably influenced by appearance. It is undoubtably true that many mongrels are very appealing as puppies and that many remain so, but the possibility always exists, in a dog of unknown parentage, of a small, appealing puppy growing into a large and very plain dog.

A mongrel is popularly supposed to be more hardy, more intelligent and free from all the weaknesses which are said to be brought about by 'in-breeding', but let us examine these suppositions. A bitch roaming the streets when in season will mate to the dog whose pugnacity is such as to keep other suitors at bay and whose size makes him able to mate her quickly and easily. In the course of time her litter will themselves be roaming the streets and the dogs to which they are most likely to mate are precisely those whose pugnacity and size made them suitable mates for their mother, in other words they are likely to mate with their own father or brother. A dog roaming the streets and mating bitches indiscriminately is likely to carry disease from bitch to bitch, and is unlikely to be in a condition to sire healthy litters. It is very unlikely that the resultant litter will be reared in such a way as to produce strong and vigorous puppies. Of course we would not claim that all pedigree dogs are the product of suitable matings or that

This Soft-Coated Wheaten Terrier and her pup, owned by Mrs E. L. Gottfeld, illustrate just how much difference, even with pedigree dogs, there may be between a young puppy and an adult.

they are all reared under ideal conditions, but the probability is that a deliberate mating will be the product of careful thought and that the resultant pups, if only because they are potentially valuable, are likely to be well reared.

It is commonly supposed that mongrels are more resistant to disease than pedigree dogs, due to a condition known among geneticists as hybrid vigour and that they do not suffer from the same hereditary conditions which afflict some pedigree breeds. It may in fact be true that some mongrels are more vigorous than a more pampered pedigree dog but it is also true that a pedigree dog which is still pursuing his original purpose in life is likely to withstand a greater degree of adversity. From our own experience we have seen how dreadfully a terrier can be punished by a fox or badger and yet quickly recover, whereas less hardy animals would certainly have died, and one has only to think of the work that is undertaken by some of the Arctic breeds of dogs and appreciate the conditions under which they live to realize that many pedigree dogs are innured to extreme hardship which no mongrel could ever tolerate.

Perhaps the simplest and most convenient way of deciding what breed of dog to get, is to attend a dog show. In the United Kingdom well over two thousand shows licensed by the Kennel Club are held annually; their location and details of the breeds for which they cater are regularly advertised in the dog magazines, and the larger Championship Shows are often widely publicized in other places. At the larger dog shows there will be an opportunity to see a great number of breeds represented by good quality dogs and you will find that exhibitors will be pleased to talk about their particular breeds. Remember though that their reaction is that of enthusiasts who have devoted a great deal of time and thought to their particular breed, as a consequence of which any disadvantages which the breed may have may tend to be ignored. However very few people will deliberately set out to mislead you and so with the use of a little common sense it should not be too difficult to decide which breed of dog appeals to you most or, at least, to narrow your choice down to a handful of breeds. It is from such people or from reading books devoted to the breed which has captured your interest, that detailed knowledge of breeds must come; there are well over 100 breeds of dog and it is obviously not possible for us to deal in detail with each breed. In the United Kingdom the breeds which are recognized by the Kennel Club are divided into six groups, Hound, Gundog, Toy, Working, Utility and Terrier, and although some breeds may appear in different groups the same basic grouping is used in other countries. Even within each group there is a fairly wide difference of type, size and so on, so that it is not possible to generalize about the group, but perhaps we can help by isolating the basic differences between the breeds in each group.

GROUP 1: HOUNDS

Afghan Hound. An intelligent dog, of aristocratic appearance, who is usually quiet, reserved with strangers and an excellent and hardy companion. He was bred to hunt leopards, and is seen at his best moving at speed across open country; he enjoys extensive exercise. His coat is thick, fine and silky and needs plenty of regular grooming. Height: 25 to 29 in. (Height is measured to the shoulder throughout this chapter.)

Basenji. A smart, clean intelligent dog, with a short and easily kept silky coat. He almost never barks but can be an accomplished howler. Height: 16 to 17 in. Weight: 21 to 24 lb.

Basset Hound. A placid breed which enjoys an outdoor life and, like most pack hounds, is exceptionally robust. His close, hard coat is easily maintained. Height: 12 to 13 in. Weight: 40 to 50 lb.

Beagle. Another pack hound who is consequently sometimes more interested in hunting than in obedience; in other respects he makes a fine pet, though he is capable of being very noisy. Height: 15 to 16 in.

Bloodhound. A large, short-coated dog which needs a lot of exercise and enjoys using his legendary hunting powers. James Thurber complained that, when he called his Bloodhound, it was less interested in where he was than in how he got there. Height: 23 to 27 in. Weight: 80 to 110 lb.

Borzoi. A sensitive sight hound capable of great affection but who himself needs great affection. His coat is long and silky and is capable of protecting him against severe cold. Height: 27 to 29 in. plus.

Dachshund. The smallest and most popular hound; there are two sizes of Dachshund, standard and miniature, and three coats, smooth, wire-haired and long-haired. All the varieties are intelligent, lively, require little exercise and are well adapted to family life. Weight: *Standard*, around 20 lb.; *Miniature*, around 10 lb.

Deerhound (Scottish). A large, exceptionally loyal dog,

A brace of Mr and Mrs Osbourne's Deerhounds, examples of the only Scottish hound bred. These tough, hardy and alert hounds are most at home in the open spaces and are members of the gazehound family, since they hunt by sight rather than by scent.

hardy, easy to train and gentle with children. His coat is harsh and wiry and easily kept. Height: 28 to 30 in. plus. Weight: 65 (bitch) to 105 lb. (dog).

Elkhound. A brave, wilful dog, who requires knowledgeable handling if his exceptional qualities are to be fully realized. His coat is thick and dense, capable of withstanding an arctic winter. Height: 18 to 21 in. Weight: 43 to 50 lb.

Finnish Spitz. A small, red dog with a long erect coat which he keeps scrupulously clean. Like most Spitz breeds he needs to be well trained. Height: 15 to 19 in.

Foxhound (English). Seldom found as a pet in Britain, where he is kept as a pack animal. Characteristics similar to the Beagle. Size is dependent on the type of terrain hunted; height usually 23 to 25 in. – the largest of the pack hounds.

Greyhound. An ancient breed, happiest when using his exceptional speed in the chase. Height: 27 to 30 in.

Harrier. In size midway between a Beagle and a Foxhound, with whom he shares many characteristics. Height: 19 to 21 in. Weight: 40 to 50 lb.

Irish Wolfhound. A gentle giant, dependable, dignified and brave. His size suits him better for country than town. Height: 28 to 34 in. Weight: 100 to 120 lb.

Otter Hound. Bred in the U.K. purely for work, at which he excels. Height: 24 to 26 in. Weight: 55 to 65 lb.

Rhodesian Ridgeback. A reserved, one-man dog who needs to be well trained and handled. His coat is short, sleek and dense with a characteristic dorsal ridge. Height: 24 to 27 in. Weight: 70 to 85 lb.

Saluki. An ancient aristocrat, reserved but a good companion. Height: 23 to 28 in.

Whippet. A dog which is happiest sharing the fireside with his master; friendly and clean but needs a lot of exercise. Height: around 18 in.

GROUP 2: GUNDOGS

English Setter. An elegant, intensely friendly and quiet-natured dog with a keen sense of game. Enjoys and should receive considerable exercise. Height: 24 to 27 in. Weight: 56 to 66 lb.

Gordon Setter. More strongly made and more reserved with strangers than the English Setter. Enjoys a lot of exercise. Height: 24½ to 26 in. Weight: 56 to 65 lb.

Irish Setter. A racy dog with a kindly expression and disposition, but with a strong will of his own; some strains are inclined to be a little nervous. Height: 23 to 26 in.

Pointer. Intelligent and loyal to those who win his

affection, though reserved with strangers. He was bred to range over open country and should be given the opportunity to do so. Height: usually 21 to 25 in, but no limits laid down.

German Short-Haired Pointer. A keen short-coated working dog, strong and determined. He is very loyal and manageable to those who win his affection and generally well mannered to strangers. He needs a lot of exercise. Height: 21 to 25 in. Weight: 45 to 70 lb.

Curly-Coated Retriever. A strong, smart, upstanding dog showing activity, endurance and intelligence. He is mild tempered and affectionate but he needs a lot of exercise and his coat needs careful and constant grooming. Height: about 26 in. Weight: 70 to 80 lb.

Flat-Coated Retriever. A bright, active dog of medium size with an intelligent expression and a docile disposition.

Mr G. Farrand's German Short-haired Pointer Champion Larberry Link, retrieving from water. This is one of the most versatile of the Gundog breeds, able to tackle many aspects of a Gundog's work other than merely pointing.

Like all gundogs he is a hardy, strong dog who needs a lot of exercise. Weight: 60 to 70 lb.

Golden Retriever. An active, powerful dog with a kindly expression and disposition. Makes a reliable and cheerful companion. Height: 20 to 24 in. Weight: 55 to 70 lb.

Labrador Retriever. A strongly built, very active dog, usually with a well-mannered gentle disposition; his short coat is easily kept, but he needs a lot of exercise. Height: 21½ to 22½ in.

American Cocker Spaniel. A sturdy little dog with a merry disposition, equable in temperament and usually without timidity. He grows a profuse coat with a lot of feathering on the legs which needs constant and careful grooming. Height: 14 to 15½ in. Weight: 22 to 28 lb.

Clumber Spaniel. A heavy, massive but active dog with a thoughtful expression. He is not always a sociable chap. Height: 17 to 18 in. Weight: 45 to 75 lb.

Cocker Spaniel. The most popular of the gundogs; usually has a merry, gay disposition, although those who have capitalized on his popularity have produced some nervous and aggressive dogs. He can grow a very heavy coat which needs careful attention, as do his pendulous ears. Height: 15 to 16 in. Weight: 28 to 32 lb.

English Springer Spaniel. A compact, strong, upstanding, merry and active dog built for endurance and activity. For a gundog he makes an unusually good guard. Height: 20 in. Weight 50 lb.

Field Spaniel. Not often met with purely as a companion, but he is a well balanced, upstanding, sporting chap, built for activity and endurance and combining good looks with good functional qualities and a very docile nature. Height: 18 in. Weight: 35 to 50 lb.

Irish Water Spaniel. Unlike all the other gundogs in appearance, he is happiest in water, of any temperature, from which his thick curly coat affords him adequate protection. Not an especially sociable chap but usually gentle and loyal to his friends. Height: 20 to 23 in.

Sussex Spaniel. A massive and strongly built Spaniel,

lacking the clownish quality of some of this group. He makes an excellent house dog. Height: 15 to 16 in. Weight: 40 to 45 lb.

Welsh Springer Spaniel. A compact, strong, merry, very active dog built for endurance and hard work. He has a robust disposition and outlook on life and needs a lot of exercise. Height: 18 to 19 in. Weight: 35 to 45 lb.

Weimaraner. A keen, fearless and friendly dog with a protective and obedient nature. He tends to be self-willed and must be intelligently handled. Height: 22 to 25 in. Weight: 45 to 65 lb.

GROUP 3: TOYDOGS

Chihuahua. An alert and swift-moving little dog with a saucy expression; has either a long or a smooth coat. He occasionally exhibits a temperament which in a bigger, stronger breed would make him quite unmanageable. Weight: 2 to 6 lb.

English Toy Terrier (Black and Tan). A elegant and compact little dog, sleek and cleanly built. His short coat needs a minimum of attention. Height: 10 to 12 in. Weight: 6 to 8 lb.

Griffon. The two varieties, smooth (Petit Brabançon) and rough (Bruxellois) are intelligent and affectionate, but occasionally possess a tendency to bronchial trouble which is characteristic of all short-faced breeds. Weight: 6 to 9 lb.

Italian Greyhound. A miniature Greyhound, built on slender lines, of great elegance in shape and movement. Enjoys warmth and the company of his master. Weight: 6 to 10 lb.

Japanese Chin. A lively dainty little dog with an arrogant carriage and profuse coat. Weight: about 7 lb.

Cavalier King Charles Spaniel. An active, graceful dog, fearless and sporting in character and a very gay mover. One of the largest of the toy breeds. Weight: 10 to 18 lb.

Champion Pargeter Trillium of Ttiweh, a Cavalier King Charles Spaniel owned by Mrs A. Pitt. This breed, one of the most sturdily built of the Toy group, is a relatively recent breed which sought to re-introduce the original form of the King Charles Spaniel, an object which seems to have been fairly successfully achieved.

King Charles Spaniel. A lovable, short-faced little dog with a long, silky coat. Weight: 8 to 14 lb.

Maltese. A sweet-tempered, very intelligent little fellow with a smart, lively, alert appearance. His silky white coat needs regular and careful attention. Height: 10 in. Weight: 4 to 10 lb.

Miniature Pinscher. A sturdy, compact, elegant, smooth-coated toy dog, naturally well-groomed, proud, rigorous and alert. Height: 10 to 12 in.

Papillon. A dainty little toy dog with alert bearing and an intelligent and lively expression who demands constant attention. Height: 8 to 11 in. Weight: 3½ to 7 lb.

Pekingese. A small, thickset dog of great dignity and quality. He is intelligent and loving and needs a minimum of exercise. His profuse coat needs constant and careful grooming. He can be subject to respiratory troubles due to his short face. Weight: 7 to 12 lb.

Pomeranian. A compact little dog whose expression

exhibits great intelligence and whose activity and buoyancy is evident in his deportment. Weight: 4 to 5½ lb.

Pug. A square, cobby, thickly muscled dog with short, easily kept coat. He seems to avoid many of the troubles to which short-faced breeds are susceptible. Weight: 14 to 18 lb.

Yorkshire Terrier. A vigorous little breed with a gay almost cheeky attitude to life; grows a long coat which needs reqular grooming. Weight: up to 7 lb.

GROUP 4: NON-SPORTING (WORKING)

Alsatian (German Shepherd Dog). This popular breed probably has more rubbish written about it than any other dog; he has suffered by being too popular. A well-bred Alsatian handled by an intelligent person can be trained to do almost anything and is a joy to own. Height: 22 to 26 in.

Bearded Collie. An alert, lively, self-confident breed usually possessing a reliable temperament. His harsh coat needs regular attention. Height: 20 to 22 in.

Boxer. A bright, friendly, clean dog, intelligent and tractable. Can tend to exhibit over-exuberance, but when properly handled will prove to be a useful guard and companion. Height: 21 to 24 in. Weight: 62 to 66 lb.

Bullmastiff. A very strong, very big dog bred as a guard and an ideal deterrent to malefactors, but one which needs to be intelligently handled and trained. Height: 24 to 27 in. Weight: 90 to 130 lb.

Collie (Rough). A strong, alert dog which enjoys a robust outdoor life with plenty of exercise. He makes an instinctive guard and is usually affectionate and easily trained. His dense coat needs regular and thorough grooming. Height: 20 to 24 in. Weight: 40 to 65 lb.

Collie (Smooth). Less popular than his rough cousin, but his coat is more easily managed. Height and weight as for the Rough Collie.

Dobermann. A tough dog, bold, quick, alert and intelligent; needs plenty of exercise. Properly trained he makes a first-class guard and companion. Height: 25½ to 27 in.

Great Dane. An elegant and graceful giant, with a giant's appetite for food and exercise. Invariably even-tempered almost to the point of sloppiness. Height: 28 to 30 in. Weight: 100 to 120 lb.

Groenendael (Belgian Shepherd Dog). An up and coming breed of the shepherd dog type, but with some differences from the German Shepherd. Needs intelligent care and can be relied upon as a guard. Height: 22½ to 23½ in.

Mastiff. Combines grandeur and good nature, courage and docility. He is seldom nervous or vicious and being anxious to please is usually easily trained. A very big, heavy dog. Height: at least 30 in.

Newfoundland. A very large dog with an exceptionally gentle and docile nature; if he can be given adequate exercise he makes a most devoted companion. Height: 25 to 28 in. Weight: 110 to 150 lb.

Norwegian Buhund. Should be a brave and fearless dog, capable of the arduous work for which he was originated. The coat is close, harsh and smooth. Height: dogs about 18 in., bitches somewhat less.

Old English Sheepdog. A friendly chap, but one who enjoys a lot of exercise and whose profuse coat needs to be groomed regularly. Height: about 22 in.

Pyrenean Mountain Dog. A carefully bred Pyrenean should be gentle, docile and courageous in the protection of anything placed in his care. Height: 25 to 32 in. Weight: 90 to 125 lb.

Rottweiler. A powerful dog capable of great strength and endurance, who enjoys an outdoor life. Height: 23 to 27 in.

St Bernard. A tough dog not built for the centrally heated flat. Capable of great benevolence, dignity and intelligence. Height: the taller the better; carrying great weight and substance.

Samoyed. Intelligent, alert, full of action; usually an even-tempered dog. His dense white coat needs regular grooming. Height: 18 to 22 in.

Shetland Sheepdog. A dog which shows affection and response to his owner but is reserved with strangers; lithe and graceful, of great beauty and intelligence. His long, harsh coat needs regular and careful grooming. Height: 14½ in.

Welsh Corgi (Cardigan). Intelligent, tough, usually long lived, but can be noisy and quarrelsome. Not so popular as his cousin from Pembroke. Height: 11½ to 12 in. Weight: 20 to 26 lb.

Welsh Corgi (Pembroke). A low set, strong, sturdily built dog, alert and active. It should be remembered that in spite of his small height he is not a small dog. Height: 10 to 12 in. Weight: 18 to 24 lb.

GROUP 5: NON-SPORTING (UTILITY)

Boston Terrier. An American breed deservedly popular on both sides of the Atlantic; makes an ideal companion for the city dweller. His short face makes him susceptible to bronchial troubles. Weight: 15 to 25 lb.

Bulldog. Despite his ferocious appearance a gentle, quiet animal but one whose exaggerated physical properties sometimes make him liable to heart and bronchial troubles; his life-span is short. Weight: 50 to 55 lb.

Chow Chow. Incredibly strong, and intensely loyal with those to whom he gives his affection, but suspicious of all others. A very heavily coated breed, which suffers in excessive heat. Height: at least 18 in.

Dalmatian. An equable dog, usually good with children, but needs freedom and exercise. Healthy and clean, his short spotted coat needs little grooming. Can be absurdly affectionate. Height: 19 to 23 in. Weight: 50 to 55 lb.

French Bulldog. Similar characteristics to the Boston, and like him tends to snore dreadfully. Weight: 24 to 28 lb.

Keeshond. Not a demonstrative dog, makes an excellent guard if he is properly trained. Height: 17 to 18 in.

Lhasa Apso. His heavy, straight coat needs a lot of grooming, but he does not need excessive exercise. An affectionate breed which can be very self-willed. Height: 9 to 10 in.

Poodle. His coat, which is never shed, needs to be regularly clipped. He has suffered by becoming too popular too quickly, but those from reliable breeders are sound, intelligent and make first class companions. There are three sizes of Poodles: *Toy* (under 11 in.), *Miniature* (under 15 in.),

Mrs Beale's Schipperke, Seagarden Black Duke, resting comfortably and contentedly on his bench at a show. This small breed was originally bred as a watchdog for use on Flemish barges.

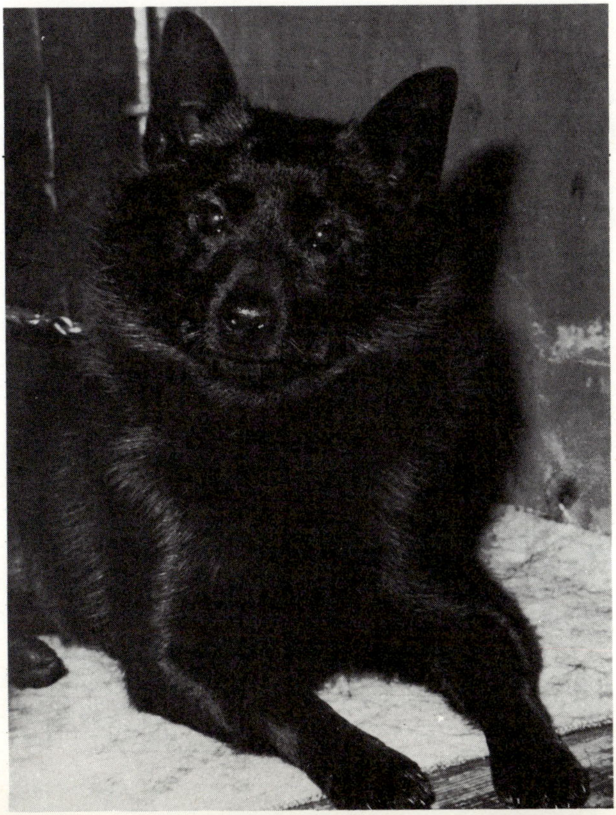

and *Standard* (15 in. and over).

Schipperke. A hardy, small dog, at home with a family. His coat is abundant and needs regular grooming. Weight: 12 to 16 lb.

Schnauzer. A sensible dog who makes a good guard; tends to be reserved with strangers. His coat is hard and wiry and needs clipping from time to time. There are three sizes of Schnauzer: *Giant* (21 to 25 in.), *Standard* (18 to 19 in.), and *Miniature* (13 to 14 in.).

Shih Tzu. An arrogant, active, hardy, ancient breed, enjoying a deserved popularity at the present time. His coat is long and dense. Height: 10½ in. Weight: 9 to 18 lb.

Tibetan Spaniel. Like the Shih Tzu and Apso an Eastern breed which is now deservedly popular. He makes a good house dog and a faithful companion. Height: up to 11 in.

Tibetan Terrier. A tough, brave breed which can make a delightful companion. Height: 13 to 16 in.

GROUP 6: TERRIERS

Airedale Terrier. The largest of the terrier group; makes an excellent guard and a good family dog, but his natural exuberance demands a lot of exercise. His dense, wiry coat needs expert trimming. Height: 22 to 24 in.

Australian Terrier. One of the smallest terriers; loyal and affectionate. His long, harsh coat is not trimmed and needs only routine grooming. Height: about 10 in. Weight: 10 to 11 lb.

Bedlington Terrier. A hardy and active breed, who enjoys nothing more than a good fight. His thick, linty coat needs regular grooming and trimming. Height: about 16 in. Weight: 18 to 23 lb.

Border Terrier. A sensible, sociable chap, hardy and game; his harsh coat needs little attention. Weight: 11½ to 15½ lb.

Bull Terrier. The gladiator of the group, a big hardy chap who needs to be intelligently handled. A fighting Bull

Mrs N. G. Hogg's Cairn Terrier Champion Myaldour My Jo Janet shown here in typical alert stance. This breed's unspoilt appearance may be one of the reasons for its growth in popularity both in the United Kingdom and abroad.

Terrier is lethal, but a well-trained one makes an excellent companion and guard. Size: There are no size limits laid down. *Miniatures* – Height: not more than 14 in. Weight: not more than 20 lb.

Cairn Terrier. Active, game, hardy and 'shaggy' in appearance; a loyal extrovert. Weight: about 14 lb.

Dandie Dinmont Terrier. Needs careful training, and his crisp coat needs regular grooming; he will then make a first-class house dog. Height: 8 to 11 in. Weight: about 18 lb.

Fox Terrier (Smooth). Gay, lively and active, a terrier with a happy, even excitable, disposition. His close, hard coat is easily kept. Height: about 15 in. Weight: 15 to 18 lb.

Fox Terrier (Wire). Alert, quick of movement and with a keen expression. His wiry coat needs to be well groomed and trimmed. Height about 15½ in. Weight: 16 to 18 lb.

Irish Terrier. A spirited, devoted terrier; given half a chance can be a ferocious fighter. His harsh, wiry coat needs to be skilfully trimmed. Height: 18 in. Weight: 25 to 27 lb.

Kerry Blue Terrier. Another spirited Irishman who needs to be carefully handled and whose coat needs to be skilfully trimmed. Height: about 18 in. Weight: 33 to 37 lb.

Lakeland Terrier. A gay, fearless, hardy breed whose harsh coat needs a fair amount of attention if he is to be kept looking presentable. Height: about 14½ in. Weight: 15 to 17 lb.

Manchester Terrier. Intelligent, quick and alert but can be noisy. His short black coat needs little attention. Height: 15 to 16 in. Weight: 17 to 18 lb.

Norfolk Terrier. A gay dog whose hard, easily kept coat and small stature make him ideally suited to town-life. Not generally a quarrelsome or snappy breed but too small to be

Mrs M. Fearfield's Griffon Champion Bowerhinton Billy Bee. This incredibly active little breed bears many resemblances to a Terrier though it forms part of the Toy group.

a good worker, though some make good ratters. Height: about 10 in.

Norwich Terrier. A slightly bigger, stronger version of the Norfolk, with pricked rather than drop ears. Height: 10 to 11 in. Weight: 11 to 12 lb.

Scottish Terrier. A very popular, heavily built breed; his coat needs a fair amount of attention. Height: 10 to 11 in. Weight: 19 to 23 lb.

Sealyham Terrier. A breed that has grown in size, and has developed a very profuse coat since it ceased to be used for work. Height: about 12 in. Weight: 18 to 20 lb.

Skye Terrier. A one-man dog, distrustful of strangers, who can at times be quarrelsome. His coat needs regular attention. Height: about 10 in. Weight: 25 lb.

Soft-Coated Wheaten Terrier. Typical Irish temperament, gay and extrovert. A breed which has much to recommend it to those who can spare the time to train it properly and to groom its abundant coat. Height: about 18 in. Weight: about 35 lb.

Staffordshire Bull Terrier. A wonderful, easily kept dog for those capable of handling him but an unholy terror for those that cannot. Height: 14 to 16 in. Weight: 24 to 38 lb.

Welsh Terrier. A trimmed breed similar to the Lakeland, with a gay volatile disposition. Deserves to be more popular. Height: about 15½ in. Weight: 20 to 21 lb.

West Highland White Terrier. A gay, almost conceited, little chap with masses of charm and vitality. Deservedly popular. Height: about 11 in. Weight: about 15 lb.

CHAPTER 2
Choosing and Buying a Puppy

Having decided upon a breed to suit your own particular circumstances you will still be faced with several other problems before you have in your possession the puppy which represents your final choice. Choosing a good puppy is, in many respects, something best learnt by experience, but in many other respects a matter of common sense. It not only varies in its detail from breed to breed but also it differs according to the purpose for which the dog is required. Naturally the same criteria would not all be applied in choosing dogs as pets, for breeding stock or for work.

First of all you must decide from whom the puppy is to be purchased. It is probably best to consult the Secretary of your chosen breed's specialist club. He will have a detailed knowledge of the breed and will be able from his list of members to advise about the availability of breeding establishments in your particular area. The names and addresses of these Secretaries can be obtained from the ruling body which controls these clubs – namely the Kennel Club relating to the actual country involved (see Appendix 1).

There are various types of people selling dogs at the present time; the main groups into which they fall are pet shops, dealers, dog homes, major breeders and casual breeders.

Dealers are usually most reliable people who can be trusted to supply you with an excellent puppy which admirably fits your needs. But there are also those who may be a little unscrupulous and who ought to be avoided at all costs. Normally the less reliable dealers will undertake to supply you with a puppy of the breed of your choice without your having viewed the puppy. This they can do either by sending off to certain regular suppliers or alternatively by buying in litters from occasional breeders who have neither the time not the inclination to sell their own puppies. Both of these methods are fraught with danger. In the first instance you will not – and the dealer will not – have seen the puppy concerned and this, no matter how cold headed you are as a business man, cannot be conducive to the best choice being made. In the second alternative method there are other dangers, in that the type of person who is willing to part with whole litters to dealers may not always be very interested in the breed or in the particular puppies concerned, and may not have reared the puppies very carefully. This type of situation is of course the exception rather than the rule and there are many dealers who are themselves breeders and who will be able to supply you with an excellent puppy to meet your requirements. As long as the transaction is entered into prudently and as long as you go into it with your eyes open, aware of the possible dangers, then you should come out of it well.

Pet shops too supply a fairly good service most of the time and in that respect they are similar to dealers. You must of course be careful not to buy on impulse or for emotional rather than for rational reasons. It would not be the first time that someone, being ruled by the heart rather than the head, had purchased a puppy, not because of its suitability but because, for emotional reasons, he felt sorry for it and was anxious to 'rescue' it from its present surroundings.

Dog homes also make a valuable contribution to the placing of stray dogs, though they do not very often deal with the sale of puppies. Most towns these days have centres where stray dogs are taken to await re-housing. Dogs from these types of establishments have the advantage of being relatively cheap and can often turn out to be most devoted pets and companions. Consideration must however be given to the problems involved. Firstly it must be asked, 'Why was the dog discarded by its original owner – was it the dog's fault or was it that of the owner?' Quite often dogs who have turned out to be in some way temperamentally anti-social will end up in such a home. Remember also the adage that an old dog cannot be taught new tricks. But again, if care is taken and if the purchase is entered into dispassionately, then satisfaction is more than likely.

Breeders can be divided into two main categories – regular breeders who breed several litters a year and probably attend shows quite often, and casual breeders who perhaps have only owned one dog and have decided to take a litter from her in order to do her some good. Regular breeders, almost invariably, are extremely interested in and enthusiastic about the particular breed concerned and if you are lucky enough to have such a person in your area you will be fortunate indeed. You will probably be able to visit the kennels and look at the puppies along with their litter mates – but do not forget to contact the breeder before descending upon him, to ensure that your proposed visit is suitably timed. Nothing will annoy him more than to be visited at feeding time or when all the dogs have been shut in for the night. The regular and large-scale breeder will possibly not give his puppies the amount of individual attention that the smaller breeder can afford to devote, but nevertheless the experience that the major breeder possesses will probably have ensured that sound and healthy stock is being sold. The small breeder too may have much experience in canine matters but may not have the resources to go in for it on a large scale. This does not mean that the quality of puppies will be any the lower – on the contrary it probably means that he will be more likely to part with a future champion simply because he has not the room to keep more dogs. More attention too can be lavished upon the litters. Provided that a modicum of care has been taken by all concerned the chances are that you will be fairly treated by this particular group of enthusiasts. As mentioned earlier, the Breed Club Secretaries will be able to help you contact such people. Though breeders do not, of course, have puppies available all the year round, they generally also have stud dogs and will therefore probably have a pretty good idea of the litters available in any one area at any one time. Another way to contact the serious

An appealing group of Miniature Schnauzers. This breed is recognized as a Terrier in the United States, but in the United Kingdom it belongs to the Utility group.

breeder is to look out for any dog shows being held in your area and to go along there to chat to a few people. Provided you do not rush up to a person trying to cope with four Labrador Retrievers, two dog bags, a folding chair and a picnic basket and trying to find an admission ticket at the same time — you will be sure to be met with a helpful reception. But do please choose the correct moment to approach people at a show.

If the breed of your choice is rare in your country then it may be necessary to import one. But first do try to contact a breeder in your own country, as an introduction to someone in the breed abroad will undoubtedly be most useful. Importing a dog can be very costly and will definitely be controlled in some way — either by the application of strict quarantine rules or by other less severe but equally important requirements such as statutory inoculations or certificates of fitness. In any instance the matter is one for the expert and once again the local ruling body in canine matters should be asked for advice.

Having decided upon the source from which to obtain your puppy it will then be time to think of the most suitable age of puppy to go for, firstly considering the best age to choose the puppy and then the best age to take it away from its litter mates. Opinions vary on these points, and, as far as choosing a good specimen of the breed is concerned, even experienced breeders have differing ideas about the optimum time of choice. Some argue that they can pick prospective champions while the puppies are still in the nest and only a few days old, while others 'run on' their dogs until they are several months old before making their final choice. However for the average person, and for practical reasons, the best age to view a litter of puppies is at about eight to twelve weeks old, though the precise age will differ from breed to breed. At this stage they will be completely independent of their mother and will have developed many of the characteristics, both physical and temperamental, which they will possess in later life. Beware of sellers who try to part with their litters at less than eight weeks old, since the chances are that they are more interested in the profit element than in selling you a good puppy. As a general rule eight to twelve weeks is the earliest time when one can look at a litter and say that certain of the puppies, as far as show points are concerned, can be ruled out. No guarantee can be given, of course, that dogs not so ruled out at this stage will eventually turn into show champions. Indeed almost all agree that it is at this stage in the process of choice that skill and experience, along with a fair amount of good luck, will contribute to a successful choice. From the point of view of training, too, a puppy of eight to twelve weeks is ideal. You can be sure that you are working with an almost blank canvas and that most of the ideas and most of the social manners that the dog picks up will have been put there by yourself, if you take him into your possession at that age. From the point of view of most breeders too, this age is the optimum time to part with their litters, as it is at this stage that the puppies begin to require even more attention than before — attention which cannot properly be given to several at once.

The sex of your puppy is another matter for consideration at this time and of course your choice will again be affected by the availability of suitable litters. However to go into the purchase transaction with an idea of which sex you would prefer is wise, and you should be aware that there will be many behavioural differences between the sexes in later adult life. In the past most pet owners preferred to have a male rather than a female simply because females come into season twice a year and have to be isolated unless the risk of conception is taken. Personal preferences and personal circumstances again enter into the decision but in recent years people have begun to show a general preference, in many breeds, for female puppies. Various factors may have contributed to this gradual change of attitude. Firstly, pets tend to be allowed to wander less now than was once the case and they are more often isolated from other dogs in gardens and other places. Also there are on the market several forms of 'keepaway', or deodorant as it ought more properly to be described, which can be applied to the bitch when she is in season. These seasons are normally three weeks in duration and occur at approximately six-monthly intervals. The deodorants distract neighbouring dogs by leaving a false scent. People are finding too that temperamentally the female is very often more suitable than the male from the point of view of the pet owner. Generalizations are dangerous but it must be admitted that, on the whole, the female tends to be less restive and altogether more relaxed than the male. One old breeder used to put it quite aptly when he said, 'Bitches are in season only twice a year but dogs are in season all year long.' While this may be an over-simplification of the problem, it does give the correct impression. The important point is to go into your purchase with all these factors in view and to weigh up all the advantages before committing yourself to one puppy or another.

Having agreed upon all these preliminary points it is wise

to look at the temperament of the litter. At this age, eight weeks, only a rough impression can be gained and it is best, as a general rule, to avoid excesses either in exuberance or in shyness. Avoid the over-excited puppy and the over-timid puppy. Perhaps a request to see the mother of the litter would be helpful so that an assessment of family temperament might be gained. However it is not wise to pay too much attention to a timid or aggressive mother who may, after all, simply be afraid for the safety of her offspring. Consider the characteristics of the breed in general and try to assess the type of temperament normal in that breed. Expect a terrier, for example, to be more lively than a gundog and expect a hound to be more independent and self-sufficient than a toy. Make your assessment on common-sense grounds, taking into consideration the fact that your attention is something which the puppies have not experienced before. Remember that even though temperament can be roughly assessed at this age, much of the dog's social behaviour in later life will be dictated by your training and by environmental factors during the period that he is in your ownership.

Perhaps the most important factors of all to consider when buying a puppy are his health and the physical picture which he presents. These show how well he has been reared, and how well he conforms to the individual breed's characteristics. Some obvious indicators of healthiness can be considered by even the most inexperienced of purchasers. Look at the puppy's eyes to make sure that they are neither watery nor full of matter, and look at his ears to make sure they are are clean. Check that his coat is well cared for, clean and not patchy. Make sure that the puppy is not coughing or sneezing and ensure that he is neither too fat nor too thin. You should be in no fear of obtaining an unhealthy puppy if you have chosen to buy from a reliable source, but use your eyes all the time. Look for any kind of unusual behaviour and beware if you spot anything which seems untoward. Ailments and ill health are dealt with more fully in Chapter Five. But remember that once the puppy has left the kennel and is in your possession, he is your responsibility. Unless you can prove that an illness had been incubating for some time, it is very unlikely that you will be able to have any legal redress upon the seller after the puppy has been in your care for a few days. If you are in any doubt at all the breeder may be willing to allow you to have a veterinary surgeon examine the puppy, though this will not often be necessary.

The final factor influencing your choice of puppy is that of type. Type is an expression used by breeders to describe the factors which go to make any one breed distinct from any other, and to assess this accurately in a puppy one must have a great deal of experience. However, reading about a

Champion Dandyhow Shady Knight, a Border Terrier owned by Mrs B. A. Sullivan. This breed, because of a continued contact with the hunting field, has maintained its original sporting characteristics and honest appearance. Because of these attributes it is the principal breed owned and bred by the authors.

specific breed and looking at adults can help a great deal in this respect, as it can in assessing general conformation and soundness. Study the movement of the puppy to consider whether or not it is typical of that of its breed. Remember that a curved bone structure, while it might be considered undesirable in some breeds, may be much sought after in other breeds. Generalizations again are dangerous and common sense and care are all that can be recommended to the new puppy owner.

Remember not to let your head be ruled by your heart; be sure that logic and common sense rather than emotion are the factors which influence your choice. Weigh up your own personal requirements and your own personal feelings before entering into any transactions and be sure that the puppy you buy is your choice and not that of the seller. In canine matters, as much as in any other kind of transaction, the legal maxim of *'caveat emptor'* is one which ought to be borne well in mind.

Finally, before completing the formalities, certain documents should change hands. You should have not only a pedigree for your new puppy but also its Kennel Club Papers. Some breeders register all their puppies before sale and will therefore have a Kennel Club Registration Card for the puppy in question. In such instances all you will require is the card and a form signed by the current owner, transferring the dog into your ownership. If on the other hand the puppy has not yet been registered you will require a form signed by the breeder giving details of the puppy's parentage and so on before registration can be effected at the Kennel Club. In either case a copy of the dog's pedigree ought to be handed over at the time of purchase. If forms, either for transfer or registration, are not readily available, they can be obtained from the Kennel Club and completed later. Confusion is however avoided by the availability of ail these documents at the time of purchase.

CHAPTER 3
Rearing a Puppy

Only when the puppy is brought home and is entirely in the care of his new owner, will it be fully realized that things in the household, and indeed in the owner's entire way of life, will have to change in the light of the many new responsibilities which have been taken on – responsibilities which are stressed deliberately throughout this book. Your care will, however, be well rewarded when the puppy has grown into adulthood. It must be remembered that the puppy is entirely dependent upon his owner for his survival, for his training and for his eventual manner and disposition towards the rest of the community. If you go into ownership with this in mind then neither will you be disappointed, nor will the puppy be let down in any way.

Feeding your Puppy. By the time that you acquire the puppy he will be completely weaned and independent of his mother and should be accustomed to the feeding programme started by the breeder. For this reason it is wise to consult the breeder in order to find out what sort of food the puppy has been used to. It is wise to try to minimize the changes as far as is possible, for the puppy will have moved from one set of surroundings into another set which is totally different. From a situation where he has been eating with at least some of his brothers and sisters in competition, he now finds himself eating alone, in different surroundings and even out of a different dish, which in itself must be something of a traumatic experience. In order to minimize the change, the breeder's instructions should be followed as closely as possible. If you do not have the opportunity of obtaining the correct provisions that day, very often he will be prepared to give you some small sample of the type of food which the puppy has been used to, in order to tide you over. It is possible that the puppy will be very reluctant to eat at first, but this need cause no great concern provided things have improved by the second day of your ownership.

Generally, the rule with a young puppy is, just as with a young child, to feed small amounts several times a day, rather than large amounts only a very few times a day. As a very rough rule of thumb, the following general guidelines can be adopted, though the exact details will vary from breed to breed as well as from dog to dog.

8 – 12 weeks: four meals per day
3 – 6 months: three meals per day
6 – 12 months: two meals per day

Some experts may favour rather more than four meals for the young puppy of eight weeks. There are very many feeding plans accepted for puppies and you will obviously have to adapt these to the requirements of your own particular breed, having taken the seller's advice carefully. However as an example of a suggested routine we offer the following programme:

Morning Meal: Cereal with warmed milk (evaporated or powdered full cream milk, diluted with water).
Noon Meal: Finely scraped meat with a little green vegetable (larger breeds can have a little dog meal well soaked in gravy before adding to the meat).
Afternoon Meal: Finely scraped meat or cooked white fish (with dog meal, as before).
Evening Meal: Principally milk meal with from time to time the addition of an egg yolk.

The above is only a basis from which to start planning a food programme and is by no means the only routine possible.

The amount to be fed obviously varies from breed to breed and only by discussing the matter with the seller and by a fair amount of sensible adjustment can the correct amounts be arrived at. The important thing is to make sure that the puppy is kept healthy by his diet, and that he becomes neither too fat nor too thin. Transition from puppy to adult feeding will be a gradual and at the same time a natural process. As such it must be played very much by ear. Eventually the meals will be cut down in number and will become slightly larger, but it must be remembered that a growing puppy will require more food per pound of body weight than will an adult dog. On occasions the puppy will leave part of his meal; if this trend becomes persistent, it is an indication that the timing and amounts of meals ought to be adjusted. It is important, however, that no sudden change in feeding habits should be made, as these can lead, especially in the short term, to upsetting the puppy and causing diarrhoea.

A varied diet is not however necessarily bad for a dog; on the contrary it can be most beneficial provided it is properly controlled and, especially in the case of young dogs, introduced gradually. As mentioned in our table above an occasional egg yolk added to milk meal can be most beneficial, though you must leave the egg white out of the puppy's diet, since uncooked egg white contains a substance which destroys one of the B vitamins. It is perfectly safe, however, in a cooked state. If you wish to feed fish you should be careful to remove all bones and to cook it and flake it finely. Eventually a small amount of vegetables may be introduced into the diet; this subject is more fully dealt with on pages 45–9. Meat should be prepared very carefully for the young dog and preferably should be fed raw and only in very small pieces. It is probably better to scrape meat for the young puppy of only eight weeks; mincing is quite acceptable about two weeks after that. Puppy meal too must be carefully selected, so that even if the puppy bolts his food there can be no danger of choking. There are many types of meal on the market; they are usually graded in size, and the instructions of the manufacturers should be carefully adhered to in order to minimize any danger to the young dog. Some types of meal require scalding with boiling water while others need only a touch of warm water. Others again are intended to be fed dry — though not to young puppies. The use of other cereals must be carefully controlled and they must never be fed in excess.

Bedding. In order to minimize the effects of other changes when the puppy enters his new home, it is wise to find out what kind of bedding was used by the breeder. If it is possible to use the same type of bedding then well and good, but if not, something similar should be obtained. A dog likes to have somewhere of his own as a bed and for that reason it is wise to obtain, right from the start, a bed which the puppy will be able to use for some time to come. Various beds and baskets are produced commercially and these can be most useful, though other, more makeshift, types of bed can be perfectly adequate provided they are easily replaced or repaired.

Right from the start the puppy must be made to realize that, whatever is chosen for the purpose, this is his bed and this is where he must sleep. At first he will undoubtedly feel strange but you must be very firm and must begin as you mean to continue. Make sure that at every opportunity he is firmly directed to his bed at the appropriate time.

The first night, indeed the first few nights, may be difficult, but perseverance on the part of the new owner is absolutely essential. The room in which the puppy is to be left must be draught free and the bed should be placed in some corner where he can have some privacy. Be sure when leaving him alone for the first night, and indeed for several nights to come, that you do not leave anything lying around which might be within his reach and might be of danger to him or in danger from him. It is best to leave him alone in the room for some time before going off to bed yourself, as he will find the situation strange and will express his worries by whining or crying. It is essential that you are firm at this stage and that when you enter the room you immediately make it clear to him that he is meant to be in bed, by lifting him and replacing him in position. This should be done with conviction and with authority, but not in any way which might frighten the puppy. The procedure will probably have to be repeated several times before you

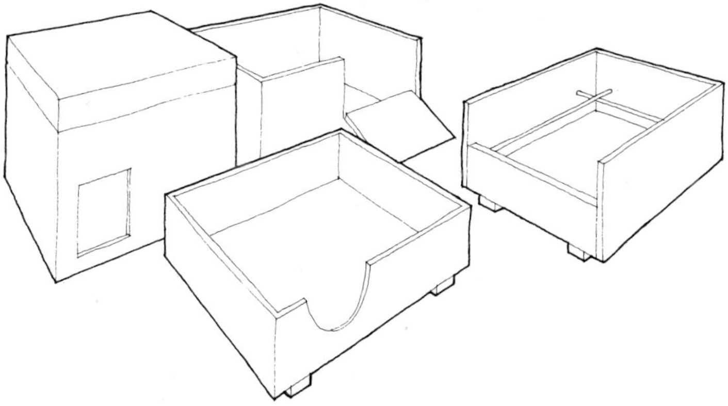

A selection of wooden dog beds which can easily be made and provide a warm, clean home for the dog.

eventually get off to bed – and it may even be tempting to employ these tactics more often still during the course of the first night. It is important however not to rush to the attention of the crying puppy too often, as eventually the habit might be adopted merely as a ruse to arouse attention. Firmness is all-important and it is absolutely essential that, no matter how much crying, wailing or whining takes place, you do not give in to the puppy or let him sleep, even on his first night, in a place where you do not intend him to sleep in the future. After a few nights, though some will be sleepless ones, these problems ought to be disappearing as the puppy gains confidence on the one hand and on the other realizes that his attempts to attract attention are being made in vain.

House training. Another immediate problem to be faced will be house training and here infinite patience may be required, though in some instances puppies do learn very quickly. Naturally in the first few days the puppy must not be allowed to have the freedom to move about the entire house, as such a freedom can naturally lead to some dire consequences and some unexpected 'surprises'.

A sensible and conscientious approach to house training usually pays off. Several methods can be used. Perhaps the most common method is to use newspaper spread about in the room where the puppy is living after he has eaten, and to watch him closely. Remember that in the very young puppy urination is automatically linked to the intake of food to the stomach as a sort of reflex action. The knowledge that this process exists can be put to good use at an early date to start habits in the young puppy which can be adapted and utilized in the house-training process. The puppy will give indications that he is about to perform by sniffing about in an agitated way and by moving round in circles. Catch him at this point, guide him to a piece of newspaper and after a short time he will learn to expect to find such a piece of newspaper and will look for it and make towards it of his own accord. Gradually the newspaper can be reduced in area and eventually moved outdoors completely. At this stage the puppy will start to indicate his wish to go outside at the appropriate moment.

Remember too that the smell of urine is also an attraction for the puppy to return to the same spot on the next occasion he wishes to relieve himself. For that reason, if accidents happen, you must be absolutely sure that no trace of urine is left and that the spot is thoroughly cleaned and disinfected, otherwise lightning will undoubtedly strike in the same place a second time. White vinegar or witch hazel applied to the spot, after cleaning with detergent and water, will take up all the stain from the urine, and help to remove the smell.

It is important to try to keep the puppy's habits as regular as possible. If you have a garden the best plan is to take him out after each meal or at the first sign of agitation. After some time the puppy, who at no time wishes, any more than his owner, to live in dirty surroundings, will become very sensible about things. At no time should the puppy be scolded for his misdemeanours, and at no time should he be scolded without knowing the reason for his owner's displeasure. The old cure of 'rubbing his nose in it' is not to be encouraged. Rather should encouragement be used as an indicator of good behaviour at the right times. Without a doubt, the problem is much greater for the owner who has no garden as there are many dangers in taking out a young, un-inoculated puppy on to the public streets. Dangers of picking up some disease at this particularly vulnerable age are great, and every effort should be made to ensure that the puppy is not exposed to these dangers until he has been adequately protected.

Inoculation. The question of inoculations is best dealt with by the expert – the veterinary surgeon. The main diseases against which vaccines are normally administered are distemper, hardpad, hepatitis, leptospirosis and rabies. Distemper and hardpad are now recognized to be varieties of the same disease and can be protected against by the one vaccine. In addition there are now multiple vaccines which will protect the puppy against all four diseases and these are well worth the extra money which they cost.

Many breeders still maintain that vaccinations are not necessary, but such diehards are becoming rare indeed these days. Most now agree that immunization is advisable, and, while it is not possible to give a one-hundred per cent guarantee of success, only a small number of those properly immunized ever contract the diseases in question.

Opinions as to the optimum age at which to administer these vaccines still vary, but immediately you acquire your new puppy you should contact your vet, who will be in a position to advise on the subject. Find out from the breeder what has been done in the way of inoculation by the time of purchase and you will then be in a position to tell your vet all he wishes to know. He will probably recommend that booster injections be given annually, and it is important to follow these instructions carefully. The small added expense at this point in a puppy's career can save a great deal of heartbreak later. (See Chapter Five.)

Worms. Nearly all puppies will have some round worms – sometimes even if the mother has been suitably dewormed before the pregnancy. By eight weeks the puppy should be quite free of worms, having been dosed perhaps twice to three times by that time. However if worms do appear to be present your vet will be pleased to administer worm tablets, which nowadays cause little more problem than does a laxative. The signs of worm infestation, apart from the presence of the worms themselves in the puppy's motion, include an extremely pot-bellied appearance (especially after meals), occasional diarrhoea and an appetite which is not consistent. The experienced breeder will normally give de-worming tablets himself, but the new pet owner is well advised to seek professional help. Having successfully eliminated any worms you should be able to see a marked improvement in the puppy's overall health, and with the improvement should come a more glossy coat and a more natural-looking puppy. (See Chapter Five.)

Teething. Like humans, dogs have, during their lifetime, two sets of teeth; the first, milk teeth, are shed at the age of about four to five months, depending upon the particular breed and indeed upon the individual dog concerned. The milk teeth number twenty-eight and will be found to be quite sharp, much to the annoyance of unsuspecting victims. At about three and a half to four months the second or permanent teeth will begin to appear, and it is at this age that the milk teeth will begin to fall out. A fair amount of pain can be attached to the process of the new teeth breaking through and can often be the cause of lack of appetite at this time. (In addition, certain breeds, particularly those which should have drop ears which lie forward close to the cheek, seem to suffer in a way which causes them to carry their ears badly and even to throw them back. There is no cause for concern, as this habit will usually disappear after the permanent teething is completed.) If the milk teeth do not come out as the second teeth are growing in, then certain irregularities in teeth formation may be caused, and it may be wise to have those milk teeth which are particularly stubborn removed. Some old-style breeders used to employ various crude methods of removing milk teeth but they must be strongly discouraged. Any problems of this sort should be referred to a vet.

Temperament. Undoubtedly temperament in the adult dog is affected by various factors, some of them hereditary and some environmental. While heredity does play quite a large part in the development of social behaviour and manners, by far the most important contributory factor in this process is that of environment. It is almost a truism that a neurotic owner will tend to have neurotic dogs and it is strange how often breeders can see this instanced in those people who buy their puppies. It is a very common happening that a family buying one puppy in a litter leaves the breeder with the impression that all will be well, and that the puppy will grow up to be a respectable member of the community, whereas the puppy's brother, purchased by another family, seems at the time of sale to be doomed to neurosis. Observations which a breeder can make at the time of sale are so often shown to be accurate at a later date. Much in a dog's make-up undoubtedly depends upon the owner's approach; it cannot be overstressed that you must start with a puppy as you mean to go on. Never allow the puppy to rule your life completely and be sure that, early in his life with you, he is left on his own for some time of the day. So often complaints are heard from doting dog owners who say that they can no longer go out to the theatre or take part in other social activities as they are tied to the house by their dog. This is obviously ludicrous and can only be the result of a wrong approach to the whole problem of dog training. Be firm with a puppy, but always be sure to reward a praiseworthy action by giving him encouragement and showing enthusiasm.

Collar and lead. Training to collar and lead is the first formal step in a puppy's education and is a task which will be easily completed. In order to accustom the puppy to the feel of a collar round the neck, it is wise to tie a loose collar or piece of string on him while he is running around the house. After a couple of days you can then fix a lead to the collar and let the puppy drag it round for a short time – always of course under close supervision. Remember that it can be extremely dangerous to allow a dog to run around out of doors with a lead attached to his collar, as this could easily be caught suddenly or become entangled while the dog tries to escape, moving in ever-decreasing circles until the lead is quite tight. Having allowed the puppy to become accustomed to the idea of having a collar and lead fixed to him, you will then be in a position to pick up the lead and try to make the puppy walk as required. First of all pick up the lead when the puppy is in some familiar place and encourage him to walk around a little. You will find that as soon as he feels the strain of the lead around his neck he will sit down. This is only to be expected – imagine your own reaction to such a situation. A few encouraging words at this stage will prove useful. A little tit-bit held out to the puppy just in front of him, so that he has to walk a few steps to take it out of your hand, will give him confidence and the lesson will become enjoyable to you both.

At first only a few short sorties into the outside world should be attempted. The sights and sounds that a puppy will meet will be overwhelming at first, but with patience and sympathy you should soon find yourself having to cope with over-enthusiasm instead of reluctance, and the puppy and you will be enjoying your walks together.

At about a year old most breeds will have established their particular temperaments and you will know how your puppy is likely to react in any set of circumstances. He should by that time know all the rights and wrongs and be able to anticipate all his owner's reactions to a given situation and the two of you should be working in harmony.

CHAPTER 4
General Care

No one should ever consider getting a dog who is not only prepared to accept complete responsibility for his well-being but also prepared to discharge these responsibilities conscientiously. Not that the conditions under which a dog should be kept, if he is to be maintained in good health, are at all complex or difficult to provide. Basically a dog needs a warm, dry bed which affords him a degree of privacy. He needs sufficient exercise to keep him fit, a well-balanced, adequate diet and regular grooming to keep him clean.

Housing. A dog may be housed either indoors or out providing that his kennel is warm, dry and clean. Some dogs, usually the smaller breeds or those which have not been developed to lead an active outdoor life, may need to be kennelled indoors or in a heated kennel out-of-doors, but the majority of dogs are quite content in an outside kennel, while just a few whose profuse coats or robust constitutions are well adapted to an outdoor life are far happier in an outside kennel than in a centrally heated house.

An outside kennel for an individual dog should be large enough to allow the dog to lie down at full stretch, to stand and turn round. It should be constructed in such a way that draughts and driving rain can be totally excluded and should have walls of sufficient thickness to insulate the interior from extremes of heat or cold. The kennel should not be sited in an exposed position or in a badly ventilated or damp one. There are firms which specialize in kennel construction and their products range from extensive rows of kennels to small individual kennels for either indoors or out. The products of the best of these firms are invariably well designed, well made and represent good value for money. It is prudent to patronize these firms rather than others whose knowledge of dogs' requirements are perhaps rather sketchy.

Indoors it may be considered that a dog needs no kennel, but in fairness to the dog he should have at least a place to which he can retire for rest or privacy, and this place should be respected, especially by the children of the household. A lot of dogs are blamed for being ill-tempered or even savage, when in order to escape the continuous and intolerable attentions of thoughtless people they have put on a 'show of force'. A great number of accidents would be avoided if every dog had a place in which his privacy was respected by every other member of the household.

The sort of kennel for a dog who lives indoors is almost infinitely variable, ranging from proper kennels, in which small dogs can be confined, to quite elaborate beds which do not look out of place even against the most luxurious background. As long as the bed is large enough for the dog to stretch out in, gives him adequate protection from draughts, and is possible to clean thoroughly, then most of the criteria of a good bed have been satisfied. For puppies, however, you must get a bed which is either robust enough to withstand their attentions while they are teething, and for this it must be very robust indeed, or easily repaired or replaced. A wooden box is ideal for even the largest breed; for smaller puppies a stout cardboard box makes an ideal bed until such times as their juvenile destructiveness begins to subside.

Apart from the bed or kennel which is to be provided, it is important that unless the garden or yard in which the dog is to be given some freedom is securely fenced he should be provided with a good-sized secure run. A dog confined to a small garden will often make his escape in order to 'go in search of love and beauty', while a bitch, particularly when she is in season, will also prospect for interesting male companions. At this time it is often as difficult to keep dogs out as it is to keep the bitch in. A wandering dog may get into more serious trouble: he may chase, injure or kill farmstock — an action for which his owner may be held liable. In defence of his animals a farmer is entitled to shoot the dog. In a town a dog may cause an accident resulting in injury to himself or to others as well as damage to property for which the owner may again be held liable. It will be seen, therefore, that in the best interests of the dog and his owner's peace of mind and financial well-being, a dog should never be allowed to wander uncontrolled. For a small dog or an inactive breed three-foot wire netting will be secure, providing that he is not able to wriggle under it or allowed to chew a hole in it. For larger or more active breeds a higher stouter fence will be required. Rather than going to the expense of fencing the entire garden it may be preferable to provide a small secure run, though of course the garden should still be adequately fenced with the more usual fencing materials. This run has the added advantage that the dog has security from the unwelcome attentions of other members of the family and can be prevented from lending a hand with the gardening. It is surprising how some breeds are able to devote their considerable energies to digging quite large holes in newly turfed lawns or in the middle of seed beds, and the urine of a bitch in season will scorch grass in a way which does not help to improve the appearance of an otherwise well-kept lawn. Runs are not difficult to make, but for those who have more important priorities they are to be had from the larger kennel manufacturers at quite reasonable prices.

It is sometimes necessary to leave a dog in boarding kennels for a while, and providing that the choice of kennel has been wisely made there need be no concern that the dog will not settle well and even enjoy his holiday and the company of other dogs. In the U.K. all boarding kennels have to be licensed, but the inspection which precedes the granting of a licence is quite a perfunctory one and gives little indication of the quality of the particular kennel, which varies between the excellent and the downright abysmal. Personal recommendation is the most reliable guide to boarding kennels, and local breeders or other dog owners should be able to recommend kennels which they have found to be satisfactory.

Exercise. The quality of exercise that a dog gets is of

greater importance than the quantity. For many dogs exercise represents not simply a way of exercising the body but also a break from an otherwise boring routine, an opportunity to meet new experiences and to develop their innate intelligence and character. Apart from the smallest toy breeds even hours of sedate walking at the end of a lead cannot hope to satisfy the body's need for exercise, and even for a small breed it does not represent a very interesting occupation. All dogs need to run free, to gallop, jump and play, and none should be denied the regular opportunity to do so.

Half an hour's exercise in rigorous play with another dog or a similar period retrieving a ball or stick provides a dog with the sort of strenuous exercise that he needs as well as great enjoyment. Of course before any dog can be allowed such freedom you must be quite sure that he is in no danger from traffic, that he will not behave aggressively towards

A well-matched brace of Japanese Chins in typically playful mood.

other animals that he might meet and that he is sufficiently obedient to return to his owner when told to do so. When these criteria have been satisfied an owner can get a lot of pleasure from seeing his dog really extend himself in the sheer enjoyment of fast movement or in seeing him playing the sort of rough games with another dog which look so furious but from which they derive such enjoyment and benefit. Few sights can compare with that of a Saluki or Greyhound in movement, or with a couple of Border Terriers really making the sparks fly. Even in quite a confined space a couple of small dogs can get adequate exercise by playing together, whereas a single dog would seldom play unless encouraged to do so.

Of course a dog's need for exercise is not constant throughout his life. Providing that a puppy is getting the necessary experience by meeting fresh situations regularly to allow his character and intelligence to develop, his need for exercise is not great, the right food and sufficient sleep being of greater importance. Older dogs too need less exercise but even so may need some encouragement before they can be persuaded to take enough. A lack of exercise in older dogs can induce constipation and heart or other muscular disorders. Exercise for older dogs should be less vigorous and they should not be encouraged to over-exert themselves, but a well-kept dog will continue to enjoy life to a very advanced age.

During a cold or wet day when the dog is being exercised he should be encouraged to keep moving and not to stand about getting chilled. All dogs of whatever age or breed should be thoroughly cleaned and dried after exercise and examined, particularly round the feet and between the pads, for any foreign substance, such as a thorn, tar or even grass seeds, which may cause subsequent trouble. The dog should be allowed to drink some water and to rest after his exertions.

Feeding. The great divergence in size, temperament, living conditions and cost between individual dogs makes it very difficult to generalize on their nutritional requirements. Apart from the obviously disparate appetites of large and small dogs, a thinly coated, active dog living out of doors and getting a lot of exercise will need appreciably more food than a dog with a more luxuriant coat, living a more sedentary domestic life. Individuals within the same breed and living under similar conditions will, because of their varying temperaments or ages, different sexes or metabolic rates, vary enormously in their needs for food. Studies made with Beagle bitches revealed a food intake which varied by almost 100%, and studies made with other breeds have shown similar results. In general it may safely be said that larger breeds require substantially less food in relation to their body weight than do smaller dogs and that puppies require about twice as much food per pound body weight as mature dogs.

Food intake of adult dogs

Total weight of dog (lb.)	Total weight of food required per day (lb.)
5	0·40
10	0·67
15	0·84
20	0·97
30	1·52
40	2·10
50	2·49
75	3·79
100	5·43

Research has shown that dogs which were allowed free access to fresh meat adjusted their food intake and maintained a constant body weight, as did dogs which were allowed free access to carefully balanced dry rations, whereas dogs which were allowed free access to canned foods became excessively fat. Dogs are basically carnivorous animals whose teeth and digestive organs are adapted to cope with large, though relatively infrequent, meals of flesh or scavenged raw meat. However in spite of this carnivorous nature their long association with man has obliged them to adapt their appetites and to accept a diet which contains some vegetable and cereal products, although it must be appreciated that this divergence from a basically carnivorous diet has been forced on them for reasons of economy or the abhorrence that their owners feel towards dogs' natural diet.

A wide range of meats are usually available which fall within an equally wide price range. It is quite feasible to feed small or toy dogs almost exclusively on good quality butcher's meat, but for larger dogs or where numbers of dogs are kept it is often necessary to consider the use of cheaper meats.

Beef and horsemeat are both readily accepted and can be given raw or cooked; a proportion of fat will help to keep the skin healthy. Offal, such as liver, kidney, paunch or tripe, heart and brains, provides a very economical and varied source of food as well as supplying many of the vitamins and other nutritional elements which are essential to keep a dog in good health. Offal for example is a source of iron and copper, which are essential to the prevention of anaemia; of vitamin A, which promotes a healthy appetite, vigorous growth and prevents respiratory infections; of the B group of vitamins, which promote normal growth, encourage lactation and fertility and maintain good muscular development; and of vitamin D, which works in association with calcium and phosphorus in promoting healthy bone and teeth development. Offal also has a mildly laxative effect which can be turned to good account with dogs which are susceptible to constipation. Rabbit and poultry too are enjoyed but great care must be taken to ensure that all bones have been removed before the dog is fed. This precaution of course applies to all food. Bones can become lodged in the dog's throat or stomach and be extremely dangerous. Cooked bones can be particularly dangerous because they are resistant to the dog's digestive juices.

Fish and eggs, preferably cooked, are useful sources of

vitamins A, B and D, and are also rich in calcium and phosphorus, which act in association with vitamin D to promote healthy bone and tooth formation. All of the above foods are sources of various proteins which are converted to body tissues and meet other metabolic needs.

It has never been shown that a dog needs any carbohydrate as such in his diet; its principle use is to provide bulk in an otherwise highly concentrated diet and as a cheap source of calories. Biscuit meal, bread (stale or baked) and other carbohydrate-rich foods should never form more than two-thirds of the dog's food intake and a proportion much lower than this is advisable if the danger of skin troubles and digestive disorders is to be avoided.

Milk and products derived from milk can have an important part to play in a dog's diet since they contain fats, calcium and phosphorus, vitamins A, B, and D.

Cereal germ meals and vegetable oils can, where meat, fish or eggs are unavailable, supply some of the dog's requirements for protein, vitamins A, B, D, and E, which is associated with reproduction and lactation. Some of the balanced dry meals which have come on to the market in recent years rely heavily on these foods with apparent success. These balanced foods are formulated in such a way that they satisfy all the essential elements in a dog's diet; but some canned foods, no matter how extravagant their advertising claims may be, rely on palatability rather than nutritional value. Of course some of the canned foods are equally good, but manufacturers are suspiciously reticent about publishing analyses of their products and so it is difficult for dog owners to differentiate between the best and those which provide bulk but little nutriment. Although commercially produced foods have an important part to play in canine nutriment and provide a convenient way of providing variety in a diet, we would suggest that for any carnivorous animal fresh meat remains the best food. Any fairly knowledgeable dog owner should be able to provide a highly satisfactory and economical diet for his dog by relying largely, if not exclusively, on a variety of fresh meats with perhaps the addition of a small proportion of meal as well as vitamin and mineral additives. Small breeds can be fed on the cheaper cuts of butcher's meat and offal, fish, eggs and milk, while the need for greater economy in feeding a larger breed can be satisfied by introducing a higher proportion of tripe, lights, melt and cheap fish such as coley.

Owners with one dog can form the basis of a perfectly acceptable diet from table scraps providing that these are supplemented with meat foods and that any sugar-rich food is avoided. Dogs' digestive systems are unable to cope with sugar, which causes skin and digestive disorders. Many dogs, given the opportunity, will enjoy a surprisingly wide range of fresh fruit and vegetables; although few dogs have any need for the vitamin C which they contain, they do make very acceptable tit-bits.

Keeping a dog in good health is very largely a matter of correct feeding, and although many pet owners provide diets which are satisfactory in quality they tend to feed far too much either by feeding too frequently or when assessing his food requirements by ignoring all the tit-bits that a dog may scrounge. To anyone with a genuine interest in dogs it is a distressing sight to see a dog of an active and robust breed so encumbered with excess fat that his breathing is impaired, his energy reduced, and he himself is able to move only with some difficulty and that neither quickly nor far. To allow a dog to get so fat makes him susceptible to many bronchial and cardiac troubles; effectively, and sometimes drastically, shortens his life; and reduces his own enjoyment of life as well as his owner's enjoyment of him. It is an insidious form of cruelty to which far too many pet owners are prone, and no one who owns a dog in such a condition has any right to think of himself as a dog owner; indeed, we would go so far as to suggest that he is not a fit person to own a dog.

Grooming. The amount of grooming that a dog needs is dependent on the type of coat that he has, on the sort of life that he lives and, to some extent at least, on the appear-

Champion Andersley Dazzler of Hadleigh, a Pomeranian owned by Mrs E. G. Anderson. This breed, though it has become one of the smallest of the Toy breeds, remains sound and hardy. The profuse coat requires more grooming than some dogs of a similar size.

ance which fashion has decreed he should have. Many of the short-coated breeds need very little grooming, even to get them ready for a show, but even so most dogs enjoy regular grooming and it gives the owner an opportunity to examine the dog thoroughly. It is therefore an important part of the regular routine in every dog's life. The short-coated breeds should be brushed with a fairly stiff bristle brush or grooming glove, and perhaps to give an extra shine to their coat finally rubbed over with a chamois cloth. Dogs with a coarser coat might need a correspondingly coarser brush and the use of a comb to free any particularly stubborn tangles. Some breeds, particularly those which grow profuse leg hair or those long-coated breeds which are built very close to the ground, need to be groomed at least daily if these tangles of hair are to be kept under control, for once they become matted the only thing that can be done is to cut off the offending hair. This takes a long time to grow again and gives the dog a very ill-kept appearance.

Many breeds require more than simple grooming to keep them looking presentable. All of the 'trimmed' terrier breeds will need expert attention to their coat at least twice a year, poodles will need trimming even more frequently and other breeds which grow profuse coats will need attention if the growth is to be kept within reasonable bounds. For the trimmed terrier breeds it is worth-while endeavouring to find someone able and willing to hand-strip the dog, that is, to remove the dead hair by plucking it out rather than by using shears or a stripping comb. It is a time-consuming and skilful operation which is more expensive than a quick once over with a pair of clippers, but it produces much better end results.

There are many so-called Poodle Parlours. Since there is nothing to stop anyone with a pair of clippers from setting

Champion Glenbeck Bannisdale Begonia, owned by Mr N. Hunt. This is a fine example of a Lakeland Terrier, one of the 'trimmed' Terrier breeds, presented ready for the show-ring. Grooming and presenting such a dog is a skilled, time-consuming task which can however be thoroughly satisfying.

themselves up in such an establishment, it is worthwhile finding one which is recommended by satisfied customers. Not all dogs take kindly to the indignity of being trimmed or stripped, particularly by a stranger, and an experienced handler can make the experience much less of a trial for them.

Bathing is a subject on which there are very different opinions. Ours is that if a dog enjoys messing about in water then, providing that he is protected from catching a chill, there is no reason why he should not be allowed to do so. Other dogs hate the very idea of getting wet and there is really no reason why they should be bathed any oftener than is absolutely necessary to keep them clean. A dog should not be washed with ordinary soap, which is difficult to rinse out of the coat and can irritate the dog's skin. There are special shampoos marketed which are specially made for dogs.

A dog's claws also need regular attention. Some dogs, particularly those which get a lot of exercise on hard surfaces, may wear their claws down and so obviate the need for much clipping, although the dew claws, where these have been retained, will always need regular attention. The claws should either be clipped with clippers especially sold for the purpose or should be filed down. Over the years dogs' teeth, particularly those whose diet contains little hard food, many accumulate deposits of tartar which should be regularly removed. Ears, too, particularly heavy pendulous ears, should be regularly inspected to make sure that they are clean and free from accumulations of wax. They should be gently wiped clean with a tissue or soft cloth, but the temptation to probe deeply into the ear should be resisted.

CHAPTER 5
Ailments and Injuries

So few dogs which are kept purely as companions or by people whose experience of dogs is limited ever achieve the glowing health and vitality which is to be seen in most working dogs, whether they be gundogs, hounds or terriers, and in the majority of show dogs. We therefore start this chapter on ailments and injuries with an apparent paradox by describing in some detail the characteristics of a healthy dog. Most dogs rely entirely on their owners for food, accommodation and exercise; since by far the greater proportion of canine ailments and injuries can be traced, directly or indirectly, to an inadequacy in one or more of these three things it is not unreasonable to say that most of the ailments and injuries from which a dog may suffer are caused by the negligence or ignorance of his owner. We appreciate that it is possible for disease to invade even the best-run establishments and that accidents can happen which are outside the control of the dog's owner; but with strict attention to cleanliness, an adequate and well-balanced diet, a reasonable amount of exercise of the right type, and a suitable, dry, draught-proof bed there is no reason why, given that the dog has been well reared, has been bred from sound stock and is, therefore, free from hereditary weaknesses, he should not enjoy the peak of condition for the greater part of his life.

A healthy dog will have a glossy, clean coat over a supple skin; his appetite, whether he is normally a voracious eater or a more choosy one, will be constant; his motions will be soft but well formed and his urine neither pungent nor excessively yellow-tinted. His eyes will be bright and the whites clear, his nose may be either hot or cold but, like his eyes, will be free from either discharge or smell, his tongue and gums will be moist, firm and, except in certain breeds, a bright pink. Both his breath and his ears will be free from any offensive smell. The average normal temperature for a

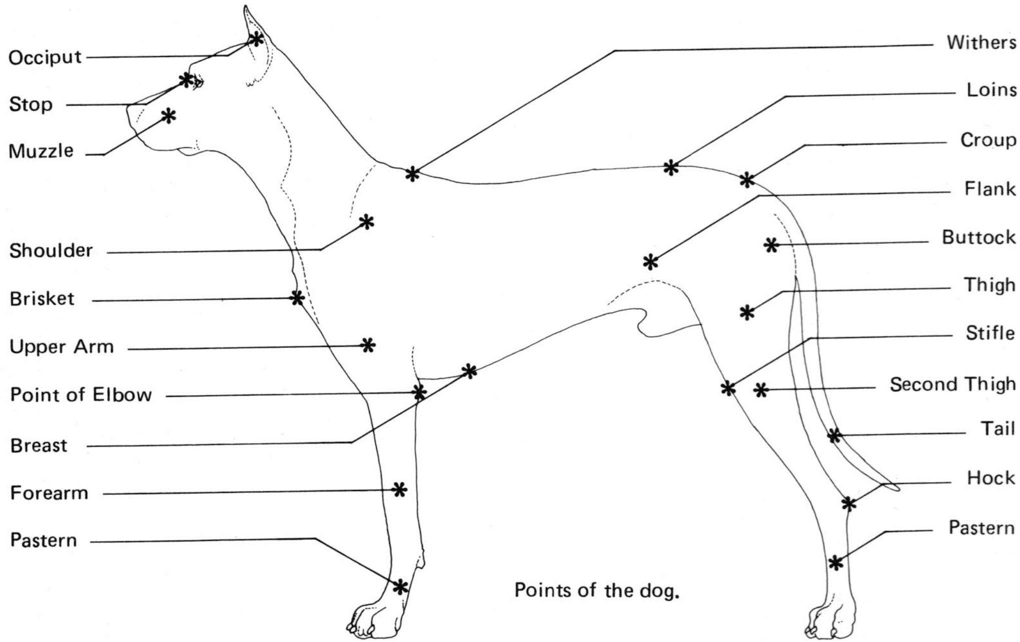

Points of the dog.

healthy dog is 101·5°F (38·6°C), but smaller dogs and bitches for whom whelping is imminent may have a normal temperature as low as 99·9°F (37·7°C), while a very large breed may normally have one of 102·0°F (38·9°C).

Dogs are creatures of habit and any departure from their normal behaviour should be viewed with great circumspection and, if the abnormal behaviour persists, should be investigated. Of course, a departure from an established pattern of behaviour may be forced on the dog by other circumstances but unless these are obvious it could mean that the dog is unwell. Prevention is better than cure. Certainly the often quite elementary knowledge and simple skills on which prevention is based are well within the scope of the average dog owner, whereas the extensive knowledge and complex skills from which cures stem are to be found only in a qualified veterinary surgeon. However some ailments which beset dogs can be cured by simple home treatment; others need expert treatment, and it is important to be able to recognize them.

Some of the more severe diseases from which dogs suffer can, to a very large extent, be avoided by the use of vaccination. The principal ones from which vaccination and subsequent periodic boosters can give immunity are contagious hepatitis, distemper and hardpad, the two forms of leptospirosis (canicola and icterohaemorrhagiae – usually called 'rat infection jaundice'), and rabies.

Contagious hepatitis. This highly contagious disease primarily attacks the liver and digestive system. Any dog which develops diarrhoea or sickness, has an excessive thirst and is obviously off colour should be suspected of having hepatitis. The dog should be isolated and veterinary assistance summoned without delay. As the disease develops the dog may get severe stomach pains and have some difficulty in breathing.

Distemper. The symptoms are similar to those of hepatitis and again the disease is highly contagious. A running nose and eyes, a harsh cough, loss of appetite, an abnormally high temperature, blood-stained diarrhoea,

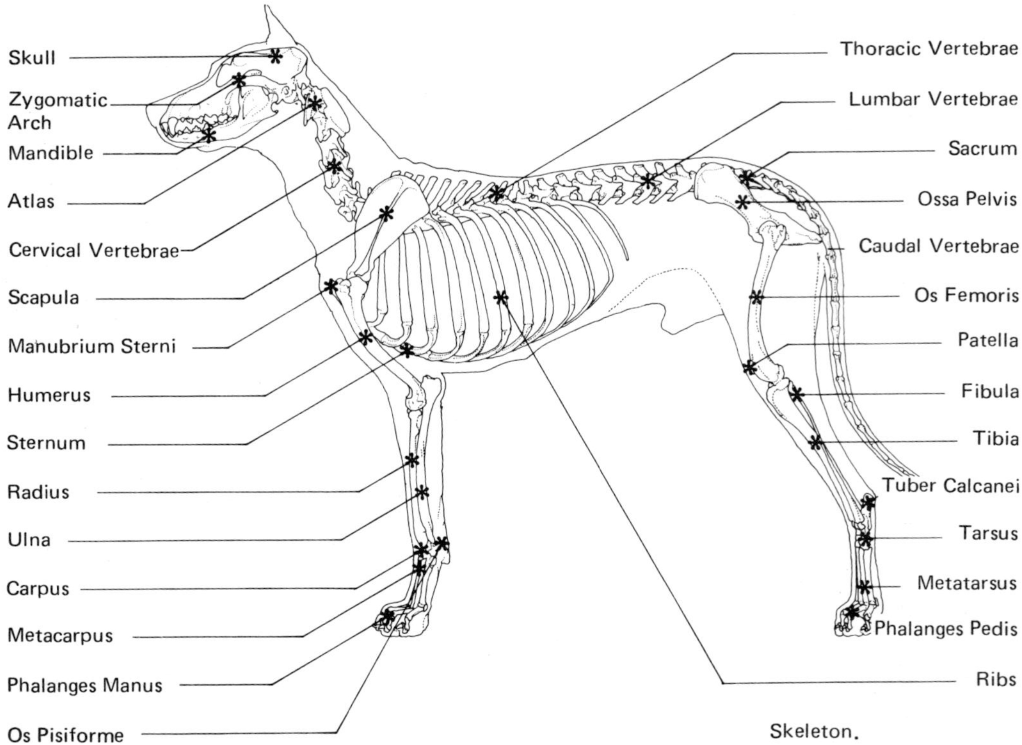
Skull
Zygomatic Arch
Mandible
Atlas
Cervical Vertebrae
Scapula
Manubrium Sterni
Humerus
Sternum
Radius
Ulna
Carpus
Metacarpus
Phalanges Manus
Os Pisiforme
Thoracic Vertebrae
Lumbar Vertebrae
Sacrum
Ossa Pelvis
Caudal Vertebrae
Os Femoris
Patella
Fibula
Tibia
Tuber Calcanei
Tarsus
Metatarsus
Phalanges Pedis
Ribs

Skeleton.

frequent urination and abdominal pimples are all signs that distemper is present. As the disease progresses the nervous system becomes infected and the chances of a complete cure are very small indeed, consequently the sooner veterinary help is summoned the better.

Leptospirosis. The two forms of leptospirosis attack the kidneys and liver and share some of the symptoms of the previous two diseases. The infection is carried in the urine of an infected rat or by another infected dog. For this reason food and water dishes, bedding, bones and toys should never be left where they can be soiled by rats. The first sign that a dog may be suffering from leptospirosis may be that his urine develops a pungent smell. The dog may develop severe abdominal pains and although he may become extremely thirsty any drink or food taken may be vomited back. It is important that the disease is recognized in its earliest stages, because the dog may suffer permanent kidney damage if it is allowed to develop. The latter stages of the disease are characterized by a jaundiced appearance of the eyes and gums and, sometimes, of the skin itself. A violently fluctuating temperature, blood-stained excrement or vomit, and apparent paralysis or extreme weakness of the hind limbs also characterize both forms of leptospirosis.

Rabies. This disease is endemic in many countries and is absent, except for very occasional isolated cases, from others, such as the United Kingdom, only because of the rigid application of strict forms of quarantine or inoculation. It is a disease from which man also suffers and which, once visual symptoms have appeared, invariably terminates in a singularly horrible death. Rabies is usually transmitted by a bite from a rabid animal who may not exhibit any symptoms but whose saliva contains the virus which quickly infects the nervous system and, eventually, the brain. The symptoms vary widely: the dog may appear very lethargic, sitting dejectedly with his mouth open and dripping saliva or he may be very alert and attack any object quite irrationally. If you have the slightest suspicion that your dog may be suffering from rabies, isolate him and call the vet immediately.

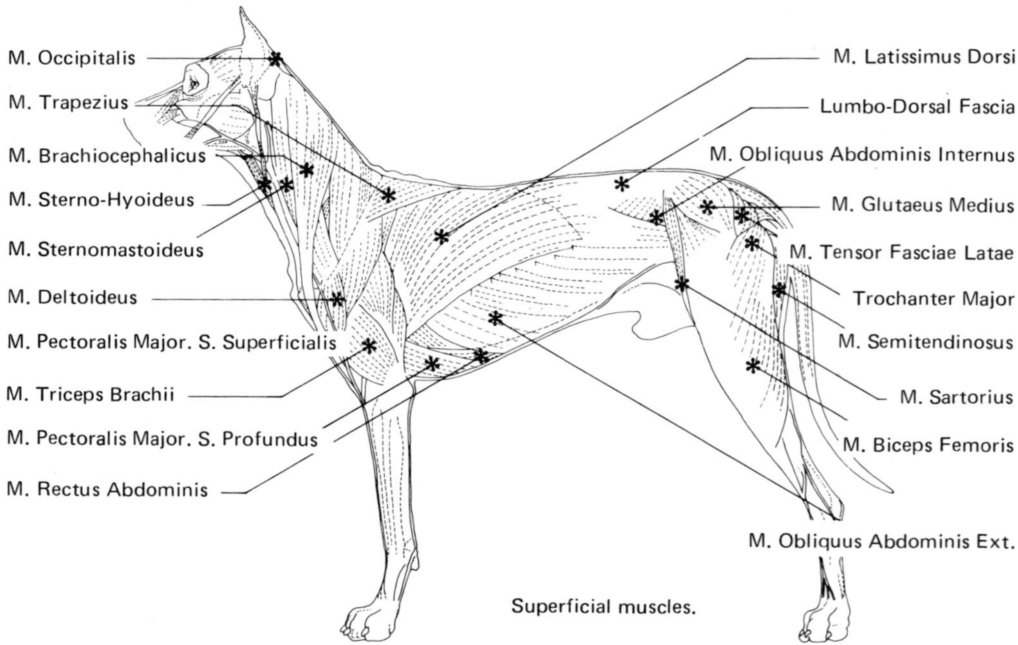

Superficial muscles.

OTHER AILMENTS

Probably the most common ailments from which dogs suffer are those resulting from injudicious feeding. Most are of a trivial nature and can quickly be set right with an adjustment to the dog's diet and some simple treatment, but all can have quite serious consequences if they are ignored. As we have stressed in the last chapter, dogs are carnivorous animals. In the highly urban society in which most of us live it is difficult to feed a dog on the variety of raw meat, offal and bones which constitute his natural diet and recourse is made to proprietary foodstuffs. There is also a tendency among many dog owners to over-feed their pets or to give them so many quite unsuitable tit-bits that the poor dog's digestive system is both over-loaded and abused. Until the dog owner is fully aware of the dietary needs of his dog he is quite likely to provoke digestive disorders resulting from an inadequacy or surfeit of some important part of the dog's diet. (See Chapter Four.)

Anal Glands. One of the minor ailments which is becoming more prevalent with the increasing reliance on soft foods, tinned or otherwise, is impacted anal glands. On each side of the dog's anal passage are two glands which become filled with excrement if the diet contains too little roughage. A swelling may appear below the anus and the dog may rub his hindquarters along the ground – an action often mistakenly taken as a symptom of worm infestation – in an attempt to relieve the irritation. If left untreated the condition can develop into abscesses and so become more serious. Treatment consists simply of applying pressure to the glands in order to empty them, a minor but skilled operation. They can be prevented from filling again by a change in the dog's diet to increase the bulk content.

Bloat. A more serious ailment and one needing rapid relief, this is more prevalent in large breeds than in small and is caused by the gas from unsuitable food, often too large biscuits or scavenged food, or by violent exercise after a heavy meal, resulting in a twisted stomach. Gas is produced in the stomach, is unable to escape, and the dog, after a period of restlessness and difficult breathing, appears to be being slowly inflated before your eyes. He should be taken to a vet as soon as possible.

Constipation. The result either of a diet which is too concentrated or of too many bones, it is prevalent in older dogs, due to the slowing down of intestinal activity, and can be induced by leaving a house-trained dog too long without an opportunity to relieve himself, or by the lack of adequate exercise. Water is absorbed from the retained faeces which then become hard and difficult for the dog to pass. Constipation can also be caused by the presence in the intestine or anal passages of some foreign body which the dog has swallowed. The treatment which relieves constipation may also facilitate the passage of this foreign body but, failing this, surgery may be necessary. It is unwise and usually quite unnecessary to try to relieve constipation by the use of laxatives unless so directed by a vet. Liver, milk, fatty meat or a nob of butter added to the dog's food all have a laxative effect and are far safer than many proprietary medicants. For older dogs an increase in the fat content of the diet, for younger dogs an increase in bulk food by the introduction of vegetables and meal to the diet and an increase in exercise, will all aid bowel movement.

Diarrhoea. Probably dogs suffer far less from constipation than they do from diarrhoea, which can result from overfeeding, particularly of milk or liver, or from a period of stress after which the dog has been allowed to become chilled. Should the condition result from overfeeding or a chill it is fairly easily corrected, but diarrhoea is often a symptom of some more serious condition and unless it clears up within a day or two expert advice should be sought. A dog suffering from diarrhoea, which is characterized by a watery and often violent expellation of excrement rather than merely watery motions, should be put on a light, bland diet of stewed chicken or chicken jelly, fish, rice, arrowroot gruel, oven-dried toast or buttermilk. The last of these, if it is available, is particularly good.

Food should be kept to a minimum and given in small frequent meals rather than larger, less frequent ones.

Enteritis. If the diarrhoea is particularly foul smelling it could be the first symptom of enteritis, which is the result of an irritation causing inflammation of the small intestine. The irritation may be caused by poisons or putrefying food, worm infestation or other foreign objects in the stomach. A dog suffering from enteritis will often take up a strange posture with his head on the ground and his abdomen raised in an attempt to alleviate the stomach pains. The dog may also develop colic, lose his appetite, vomit and have diarrhoea after trying to quench the excessive thirst which accompanies the ailment. A reluctance to move and progressive weakness develop as the condition progresses. A close watch should be kept on the temperature and motions of any dog suffering from enteritis. Should the temperature become sub-normal or the motions become blood-stained veterinary assistance should be summoned immediately. The dog should be kept warm and on a very light diet consisting of meat juices, or white of egg in warm milk to which perhaps glucose or honey has been added. If the patient is given ice cubes to lick, his thirst will be assuaged without overloading his stomach.

Gastritis. The symptoms of gastritis are very similar to and, consequently, easily confused with enteritis. It is so often the result of faulty feeding or from inflammation caused by parasites or other irritants. The dog develops an intense thirst, but even water is vomited up, his temperature rises by 2 or 3°F (1-2°C), and he often searches for a cool spot to lie in. If the patient is a strong adult food can be withheld for up to forty-eight hours to allow time for the infection to subside, but an older dog or a young puppy should not be subjected to such treatment and a bland diet should be given. Small meals of scraped beef, cheese, buttermilk, arrowroot gruel, white of egg in milk, rice or beef or chicken extract given at frequent intervals will be less inclined to overload the stomach than larger meals.

Teeth. Apart from causing digestive troubles faulty feeding can contribute to tooth decay which itself can aggravate digestive disorders or even contribute to heart disease, by allowing infection to enter the dog's system. If a dog's teeth are to remain in a healthy condition his diet should include milk or cheese, which contain a proportion of calcium, phosphorus and vitamin D. His diet should also contain sufficient hard meal or bones to keep the teeth free of tartarous accumulations. Tartar is a hard, brown deposit which, if it is allowed to develop, will slowly displace the gums, allowing access for disease and loosening the teeth. Small deposits of tartar can be removed by a mildly abrasive agent such as salt, but if the tartar is thickly encrusted it may be necessary to use a tartar scaler.

Internal Parasites. It is surprising how many owners allow their dogs to suffer as a result of the depredations of internal or external parasites which stem largely from dirty living conditions.

Worms are by far the commonest internal parasite and almost all dogs suffer from some type of worm infestation at some period of their life. Puppies are often infested with roundworm, and since an adult roundworm can be anything up to four inches in length this can be seriously debilitating for a young puppy or for a small breed of dog, particularly if the infestation is left untreated. The symptoms of roundworm or ascarid infestation are harsh, dull coat, foul-smelling breath, variable appetite and a pot-bellied appearance, particularly after meals. It is possible, but unwise, to buy proprietary medicants from a chemist or pet store; by far the safest and most satisfactory course of action is to get a reliable and safe medicant from a vet.

Hookworm is a more dangerous parasite. Hookworms feed from the blood of their victims by attaching themselves to the walls of the intestine. Symptoms of a hookworm infestation are blood-stained motions, blood-shot eyes and severe anaemia. For small, very young or elderly dogs an infestation can be very weakening and, unless satisfactorily treated, can be fatal. The treatment should only be carried out by a vet.

Tapeworms are more common in larger breeds, which for financial reasons are more likely to be fed on cheap, infected meat. Segments of the worms can become detached from the head, which remains attached to the intestinal wall and is able to grow new segments. The detached segments, which may be filled with ripe eggs, move through the intestine and as they reach the rectum can cause intense itching. To obtain relief the dog may slide its rear along the ground. Treatment should be prescribed only by a vet but is simple and effective.

Whipworms are less common internal parasites, which produce symptoms similar to those of hookworms but which may require a surgical operation for their successful elimination.

Heart-worms can sometimes be contracted by dogs which are subjected to attacks by mosquitoes; symptoms are an incessant cough, debility and difficulty in breathing. A vet can make a positive diagnosis by means of a blood test, and although treatment is protracted it is usually successful. Protection from mosquitoes, fleas, lice and ticks will reduce the chances of a dog becoming infested with heart-worms.

Apart from treating the dog to eliminate the particular internal parasite, he, his kennel mates as well as members of his owner's household should be protected from the risk of further infestation by scrupulously scrubbing out all the kennels, beds, runs and dishes with a suitable disinfectant and by keeping all at a very high standard of cleanliness.

External Parasites. Like internal parasites, the external are often a result of insanitary living conditions.

Fleas appear to be most troublesome in warm, humid weather, when they can spread rapidly from dog to dog, causing the dogs to scratch vigorously, often breaking the skin and setting up eczema. An infestation of fleas can often be controlled by the liberal use of proprietary insecticidal dusts and shampoos and by scrupulous cleanliness. However total elimination is often dependent upon identifying the source and either eliminating or avoiding it.

Lice are less active than fleas and, being smaller, are less obvious. An infestation causes intense itching and can result in severe skin trouble. Prolonged infestation by lice can cause anaemia, especially in a puppy or toy breed. The insecticidal dusts and shampoos used for fleas will also kill lice, although total elimination requires identification of the source and scrupulous attention to cleanliness in the kennels.

Ticks are particularly troublesome parasites to eliminate once they become established. Ticks are insects that look like small warts or seeds and are usually to be found on the stomach, pads, ears or in other crevices. Once they are attached to their victim they are reluctant to move and the owner may be tempted to try to pick them off. However such action will invariably leave the tick's head firmly attached to the dog and this will quickly set up a skin infection. A tick's hold can often be loosened by swabbing it with alcohol or some substance containing alcohol such as surgical spirit, whisky or, if the owner has better uses for his whisky, shaving lotion. Once loosened the tick may be carefully removed and should then be burnt. The dog's living quarters should be washed out with a suitable disinfectant to kill any eggs and the dog re-examined carefully from time to time to ensure that ticks have not re-established themselves.

Mites belong to one of four basic types: those which attack the ear, and the demodectic, follicular and sarcoptic mites which cause mange. Ear mites lead to severe irritation which causes the dog to shake his head and to scratch his ear. Inspection of the ear will reveal a dark, waxy discharge which is accompanied by a sour smell. Cure can be effected simply be carefully cleansing the ear with an oily lotion sold for the purpose, and re-infection can be prevented by keeping the ears clean.

Mange may be either demodectic or follicular or the more serious sarcoptic type. Follicular and demodectic mange may be relatively mild, appearing as little more than a reddening of the skin and some loss of hair, but may also be more severe and be accompanied by small pimples and intense irritation. The condition cannot be positively diagnosed without the microscopic examination of skin scrapings, and a complete cure is quite difficult. Shampoos, injections and both internal and external medication may all be tried by a vet to arrest the spread of the disease and relieve the irritation. If the condition responds quickly to treatment hair will grow again on the infected places but may not do so if the condition is allowed to develop. The condition is highly contagious and scrupulous attention to cleanliness must be observed to prevent the mange from spreading and to prevent re-infestation.

Sarcoptic mange causes even more intense irritation and is often associated with bloody lesions which form the scabs characteristic of this condition. The coat will be lost and the infection may spread rapidly and the dog develop a strong cheesy odour. However, like demodectic mange, positive diagnosis can only be made after microscopic examination of skin scrapings and treatment should always be in the hands of a skilled vet. This condition is highly contagious to both dogs and human beings, and so the observation of the rule of absolute cleanliness is, if possible, even more vital than usual.

Respiratory diseases. Just as dirty living conditions can induce or aggravate a variety of afflictions so also can damp, cold or draughty kennels cause unnecessary sickness and suffering for their occupants. The root cause of the majority of respiratory troubles such as asthma, pneumonia, pleurisy, colds and kennel coughs, as well as rheumatism and arthritis, often lies in the conditions under which the unfortunate animals are obliged to live.

Asthma is a disease to which dogs that have been allowed to become obese and short-faced breeds such as bulldogs are particularly prone. The condition, which is incurable, is characterized by laboured breathing, frequent bouts of coughing and loss of weight. It can be alleviated by strict adherence to a correct diet, by avoiding changes of temperature and damp conditions. The sufferer should be housed in warm, dry, well-ventilated conditions.

Pneumonia can result from shock following an accident or as a secondary infection to some other respiratory disorder. It is particularly common amongst dogs which have been weakened by neglect, age, injudicious feeding or by being obliged to live in damp conditions. It is characterized by a temperature as high as 106° F (41° C), laboured breathing and a sharp, painful cough. Treatment is similar to that for asthma, with the addition of penicillin or other antibiotics administered by a vet.

Pleurisy has similar causes to those of pneumonia, with which it is often associated. It is an inflammation of the lungs which can cause the lung cavity to fill with fluid. It is often accompanied by a harsh, painful cough and should be treated in the same way as pneumonia.

The common cold does not attack dogs as it does humans, but they can contract an infection of the upper respiratory tract which causes a wheezing cough and which can be a preliminary to pneumonia.

Rheumatism and *arthritis* are painful conditions which attack the joints and are often accompanied by swelling. Both conditions are aggravated by damp living conditions and, once contracted, are largely incurable although they can be alleviated by satisfactory housing.

INJURIES

Apart from troubles which may stem from some infection there are those which are the result of accidents, some resulting in only minor injuries requiring only simple first-aid treatment but some more serious injuries which need urgent and expert treatment. A dog which has been injured may need to be restrained to prevent him aggravating the injury or from causing injury to others. He should be restrained firmly but gently and an effort made to calm and soothe him. If the dog is unconscious he should be moved to a place of safety but care should be taken when moving him to ensure that his position is changed as little as possible so that broken bones, internal bleeding or other serious injuries are not aggravated.

If the dog's lips, gums or other parts of exposed skin become pallid, the pulse rapid and weak, respiration shallow and the temperature sub-normal the dog may well be suffering from an internal haemorrhage and should be taken to a vet as quickly as possible. In cases of severe external bleeding an effort should be made to reduce or control the bleeding by the application of pressure to the wound by means of a gauze pad and bandage. Once bleeding has been stopped the wound should be treated with an antiseptic and protected by a clean bandage. Wounds which do not penetrate the skin, bruises or contusions may swell alarmingly and can be extremely painful. The wound should be carefully cleaned with a reliable antiseptic and a cold compress applied to reduce the swelling and relieve the pain.

A dog which has been involved in an accident may have broken bones. He should, as far as possible, be prevented from moving and should be quickly taken to a vet. It is sometimes very difficult in the tense period following an accident to decide whether the dog has broken bones, a sprained joint, a dislocation or a torn muscle or tendon; however an experienced vet will be able to make an accurate diagnosis and recommend the appropriate treatment.

Apart from the injuries received in an accident a dog may also be in a state of shock, the symptoms of which are very similar to those of internal bleeding. Efforts should be made to raise the dog's temperature and he should be kept as quiet as possible. If the dog is fully conscious he may be given a drink of warm milk and glucose or soup.

Poisoning. Accidental poisoning consequent on the dog swallowing germicides, any of many of the preparations used to control pests or weeds in the garden, detergents or corrosives, should be treated as quickly as possible. As long as there is no possibility that the dog has had access to strychnine, often used to make poison bait for moles and some other vermin, he should be given a weak solution of hydrogen peroxide, common salt or mustard powder, or a small piece of washing soda should be pushed down his throat. However in all cases of poisoning the dog should be treated as quickly as possible by a vet, who will be able to prescribe specific remedies for the particular poison. Home treatment should only be used to keep the dog alive until a vet can be consulted and to alleviate the more distressing symptoms.

Burns. Burns or scalds from heat or chemicals should be carefully cleaned after cutting the hair away from the affected part. The burn should be treated with a suitable ointment and dressed to prevent infection. If the burns are extensive or in any way severe a vet should be consulted.

In general the first aid that may be applied to a dog is very similar to that which is suitable for human beings suffering from similar injuries but should there be any doubt expert assistance should be sought as quickly as possible.

CHAPTER 6

Training and Obedience

The training of dogs for particular jobs of work is a highly specialized undertaking, indeed we do not think that it is setting the skills of the best trainers at too high a level to suggest that it is a highly specialized art. Gun dogs, herding dogs of various kinds, guard dogs and coursing and racing dogs not only need specialized training but are specialized breeds bred for a particular purpose and developed over many years and generations with that purpose in mind. It is not within the scope of this book to try to describe these specialized forms of training but we must assume that the principal ambition of our readers is to train their dog to become a civilized companion which will not disgrace them in any company or situation.

Formal training should not be started in earnest before the puppy is about five months old and those breeds which derive from shepherd dogs of various kinds may not be ready for teaching even at this age. All dogs develop at different rates; if training is begun too soon a dog's spirit may be broken or, at best, he will become confused and his progress slow and halting. During his puppyhood the dog will have learned the difference between right and wrong and to respect his owner's wishes. He will be anxious to please and should be introduced into his training programme in such a way that he enjoys his lessons and the trust that has been established between him and his owner is not broken. Short lessons, perhaps of no more than ten minutes' duration, during which the puppy is praised lavishly for his successes and reprimanded only gently for any transgressions, are quite adequate at the beginning.

Various Canine Societies and Training Clubs run classes which can help considerably in training a young or recalcitrant dog by providing an atmosphere in which the dog is anxious to learn and by making expert advice and encouragement available to the novice owner. The Kennel

Club will be able to provide you with the addresses of local organizations which run such classes. When you have located a class and made the decision to attend you would be well advised to go along initially without your dog, so that your own lack of confidence will not unnerve your dog. Even when you do introduce your dog to the classes it is as well simply to allow him to become accustomed to the new experience and to enjoy the company of other dogs. When you are quite sure that he has settled and is no longer disturbed by the new situation, it is time to introduce him to the lessons that he has to learn.

If you feel that training a dog to the standard to which you aspire is beyond your capabilities or will take more time than you have at your disposal you may consider the possibility of placing him in the hands of a professional trainer. Such people are however few and far between and the services of the best trainers are in great demand, so you will have to make prior bookings if you want your dog to be trained by one of the top trainers. It is sometimes almost as difficult to get a dog into one of the best training establishments as it is to get a boy into Eton. The advantage of placing your dog in the hands of a professional is that he will be trained more quickly and to a higher standard than you could easily achieve. The disadvantages are that you will not have the satisfaction of training him yourself or of building on the bond which should exist between dog and master. The principal disadvantage however is that training should ideally be a mutual experience when both parties in the association are learning equally. The work of a good trainer can easily be undone by an owner who is unsympathetic or unable to handle his dog. Indeed we would go further and suggest that all too often owners are in greater need of training than their dogs.

The only piece of equipment that you will need is a choke chain or a leather or webbing choke which will, while you are training your dog, replace the collar that he may normally wear. The choke should be put on the dog with the end to which the lead is attached passing over the dog's throat, so that when pressure is released the choke automatically goes slack. The purpose of the choke is to enable the handler to give the dog a sharp reminder of his owner's wishes. The choke will normally be quite slack and will remain so while the dog is behaving himself. When the dog misbehaves a sharp tug on the choke will momentarily tighten the choke and bring his attention back to the job in hand.

Probably the first lesson that you will undertake will be to teach your dog to walk quietly at heel. The dog should be on your left and his lead should go across the front of your body and be held firmly in your right hand. Your left hand should touch the lead only occasionally for additional control. Tell the dog, in a firm tone of voice, to 'Heel', at the same time giving a tug on the lead and walking in the desired direction. The dog will accompany you. While he is walking quietly at your left heel all is well, but should he or his attention begin to wander another tug on the choke accompanied by the order 'Heel' will bring him back. Eventually the order alone will suffice to bring the dog to heel and less use can be made of the choke. When the dog has completed an exercise satisfactorily he should be praised effusively and made to understand that he has done something that has pleased you. Most dogs are anxious to please and will respond to this treatment more readily than to coercion. Remember to keep the lesson short and to do your utmost to ensure that both you and your dog enjoy the lessons.

The next basic lesson to attempt is to teach your dog to sit on command. Again your dog should be on your left with the lead held in your right hand and across your body. Give him the command, 'Sit', press his hindquarters firmly down and give a gentle upward tug on his choke. Once he is sitting praise him, but when he attempts to move repeat the order and tug the lead. When he will sit on command the exercise can be combined with the heel exercise. The dog should walk at heel for a while and then be told to sit, after which the heel exercise can be recommenced. Eventually the dog will walk automatically at heel and will anticipate the order to sit when his owner stops.

When these two lessons have been thoroughly absorbed the dog can be taught to stay in the 'Sit' position. The dog should be made to sit, the owner should then move slowly away from him to the extent of the lead, after giving the order, 'Stay'. If the dog attempts to move the order should be firmly repeated. The exercise should be repeated until the owner is able to stand at a distance from the dog without his attempting to move. When this exercise has been learned the owner can begin to teach the dog to stay while the owner moves out of sight. For this he will find it helpful to have some assistance to let him know if the dog moves. This exercise is not an easy one to teach a dog, because it goes against his desire to be close to his master; but with patience, praise for his success and a desire to please on the part of the dog it should not be beyond every owner and his dog.

Another basic exercise which it is important that the dog should thoroughly learn is to come to order. Most dogs will return to their owners when their name is called, particularly if they are to be rewarded in some way, either by praise or with a tit-bit. When the exercise is formalized as part of a training programme the order 'Come' should be substituted for the dog's name and eventually the exercise combined with the Stay.

Most dogs who frequently accompany their owners are expected to travel a lot by car and it is essential that they should do so without making a nuisance of themselves to other occupants of the car and, above all, without distracting or nudging the driver. The secret is to start early in the dog's life and to insist firmly that the dog travels in his appointed place, sitting or lying on the floor or at the back of an estate car.

Some dogs can become inveterate barkers unless the tendency is checked firmly and quickly. To do this it is essential to understand why the dog is barking. It may be as a result of fear or insecurity, it could be a display of temper

An impressive line of dogs at the 'sit', whose owners have moved away from them. A well-trained dog is happier and more contented than an undisciplined dog and the satisfaction gained in training and owning such a dog can be most rewarding.

or the dog might simply be trying to draw attention to itself. Until the reason for the barking has been correctly identified it will not be easy to control it, but once it has been identified it can be dealt with by removing its cause and by making it quite clear to the dog that he must not bark.

If you find that your dog is easily trained and that you enjoy training him you may be encouraged to try him in obedience competitions. Basically there are five progressively more difficult tests which are set by the Kennel Club. These are tests for Beginners and Novices, Tests A, B and C.

The Beginners' test comprises heel on a lead, heel free, sit for one minute with the handler in sight, down for two minutes with the handler in sight, recall from sit or down, and retrieve an article provided by the handler.

For Novices the exercises are the same except that the article to be retrieved must be a dumb-bell. Except in the sit and down tests the handler is not penalized for giving encouragement or extra commands to his dog.

Test A comprises heel on a lead, heel free, sit for one minute with the handler in sight, recall from sit or down while the handler is walking away, retrieve a dumb-bell, down for five minutes with the handler out of sight, and an exercise of scent discrimination.

In Test B the dog must heel free, sit for two minutes with the handler out of sight, be sent away, dropped and recalled, retrieve a dumb-bell, stand for one minute at a distance from the handler, down for ten minutes with the handler out of sight, and successfully complete a scent discrimination test.

Test C is the most demanding of the five tests and comprises free heel work at fast and slow paces, coupled with a sit and down exercise, sit for two minutes out of sight of the handler, send away, drop and recall, retrieve, stay at the down for ten minutes with the handler out of sight, completion of a scent discrimination test and obedience to the commands to sit, stand and down at a distance from the handler.

CHAPTER 7
Showing

The showing of dogs is a hobby which has grown in the last fifty years or so. Britain lays claim to being the first country to have gone in for dog showing: the earliest records of formal shows date back to just over one hundred years ago, and Newcastle on Tyne claims to have been the site of the first formal dog show in the 1850s. Initially dog shows started with the idea of breeders wishing to compete against other breeders to show just how well they were performing in their breeding programmes and to measure their success against the success of others. Naturally it was in the sphere of working dogs – whether for bull baiting, rat catching, hunting, retrieving or other forms of working – that dogs were mostly bred and shown in the early days. Gradually, over the years, the showing aspect has become more and more important and has grown in stature as compared with the working aspect of many breeds. This trend, we feel, has not in many cases been for the genuine betterment of the breeds involved. We like to think, however, that dog showing is still only one part of the wider hobby of dog breeding, and it is certainly with that kind of attitude that we would like to see newcomers entering the dog fancy.

The growth of dog showing over the past thirty to forty years has been quite phenomenal, as indeed has been the growth in the dog population. For this reason a good deal more commercialism has entered into the 'dog game' and this has not always been to the good of the fancy in general or indeed to the particular breeds so affected. We do not entirely agree with the wish and belief which constantly appear to be present in the breed enthusiast that his or her breed should become more and more popular. Better public relations for the dog fancy is undoubtedly something which we would favour, and so long as this recognizes that a breed's growth in popularity should be measured and

Champion Gisbourne Inca, a Standard Wire-haired Dachshund owned by Mrs B. Farrand. This particular dog has won 60 Kennel Club Challenge Certificates, more than any other dog of any other breed, and at the age of nine is still winning in the United Kingdom show-rings.

controlled then it can do nothing but good. Publicity for other and commercial reasons only is not something which we consider any breeder should think desirable.

There is no question that there are some cranks and eccentrics in the dog-showing fraternity, but then this is very much the case in almost all walks of life. The vast majority of the show going group are however perfectly well-balanced citizens who have a most absorbing and indeed rewarding hobby. We will concede though, that the dog-showing 'bug', when it takes hold, often leads those so afflicted to a position where they will do things which other slightly more normal folk would consider at best to be a bit odd or at worst completely round the bend. Setting off in the evening for a night's driving, spending the next day in some muddy field in the pouring rain, only to return perhaps five hundred miles the next night, is not normal in the eyes of everyone — but to the dog-showing fancy nothing could seem more natural.

There is no doubt that winning is important to us all in the dog game, but it must never become the be-all and end-all in itself, or we would be better out of the fancy, both for our own good and for that of the breed in which we have taken an interest. Winning must be important only to show that we have succeeded in breeding or choosing a better dog than the next man, and the long-term breeding programme should be given more attention than the short-term or quick win. Provided that dog-showing is kept in perspective and is not considered as an end in itself, then its future is assured. It is in recognition of this philosophy that many of the ruling bodies which control canine matters throughout the world stipulate that dogs and bitches cannot be shown once they have been sterilized, unless it has been proved that they have produced offspring at some time in the past and have registered progeny. This is one

A view of the Poodle rings at the Championship Dog Show at Windsor. The famous castle in the background makes this one of the most picturesque settings in Britain for such an event.

very sensible rule which ought to be preserved in the dog-showing field, and ought to be maintained in the context of the dog-breeding hobby in general.

This general attitude to showing and winning is essential to a full enjoyment of the hobby. Naturally, since judgment of any kind is a subjective exercise, you should never go to a show confident of winning. You should however always be confident of the excellence of your dog; otherwise why take him to the show at all? With a lack of enthusiasm for him you will undoubtedly fail to do him justice in the show ring. Remember that lack of success at the first few shows which you attend does not necessarily mean that you should stop showing your dog – far from it. Firstly, on any particular day your dog may not be looking at his best; he may be in poorer coat than required, too fat or just feeling out of sorts on that particular occasion. This does not mean that some time later, with more experience of the show ring, with greater confidence on your part, and with better presentation and condition, the same dog may not be winning well for you. Remember that if the one dog triumphed at every show there would very shortly be no shows at all, and remember too that different judges have different opinions and that they will lay different emphases on various breed points. If at first you don't succeed, then try again!

Before a dog can be shown at a properly recognized Kennel Club Show it must itself be registered at the Kennel Club, as described in Chapter Two. Registration regulations differ from country to country. Lists of rules and registration forms can be obtained from the relevant Kennel Clubs listed in Appendix 1. Generally speaking, in order to be able to register a dog you must have a note of the names and registration numbers of his parents and a form signed by his breeder. In certain instances the breeder will have a registered prefix, which means that he and he alone will be able to use this word in the registered names of all his dogs; he will most likely have given the dog a registered name before he comes into your ownership. All that will be

Mr Fred Sills, one of Britain's professional handlers, preparing Mrs J. Taylor's West Highland White Terrier Champion Checkbar Tommy Quite Right for the show-ring. Professional handlers are consummate artists in presenting and handling their clients' dogs to the best possible advantage.

required to be done in such an instance will be to have the ownership of the dog transferred in the Kennel Club Register to the new owner. It must be remembered that the purpose of Kennel Club Registrations is to control the breeding of stock and to ensure that each dog of a breed is distinguishable from every other dog of that breed, so that pedigrees can be both meaningful and reliable and so that breeding programmes can be carried out in a properly controlled way. For that reason, if you are registering a dog you should try to choose a distinctive name as, generally speaking, no two dogs in any one breed can have the same name unless a certain amount of time has elapsed between one registration and the next.

Normally dogs can be shown from the age of six months. From then until they are twelve months old they will be classed as puppies for the purpose of showing. At around five months formal training for show can begin; this may take several forms. Perhaps the best way to find out about dog showing is to attend a few dog shows first of all as a visitor. But there do exist in many areas training classes designed not only to give help to the prospective exhibitor but also to give experience to the young and inexperienced exhibit as well. Much about showing techniques can be learnt by the exhibitor and much about show atmosphere can be learnt by the exhibit. Basically what will be required of an exhibit at the non-obedience type of show will be that he should be prepared to be handled by the judge, that he should behave in a controlled way towards other exhibits and that he should be able to move up and down the show ring in a civilized manner. Apart from the normal training one or two other points ought to be considered.

Firstly the sight of many other dogs for the first time may mean that your dog will be somewhat overawed by the occasion or that he may possibly become slightly over-enthusiastic. Training clubs can be useful indeed in over-coming this problem and really the only other way to give a dog experience of this sort is to take him to a show. This you cannot do unless he is previously entered for it. Ring procedure at shows varies from country to country, but in the United Kingdom it is normal that all dogs in a class are first moved round the ring together; next each each dog is examined individually and moved individually; the judge then moves along the row of dogs assessing them further before pulling his various winners into the centre of the ring. For this reason you should be sure that your dog is used to being handled by strangers. This can best be started

A view of some of the rings at Cruft's Dog Show, with the Pyrenean Mountain Dogs being judged in the foreground.

by allowing friends, indeed by encouraging them, to handle your dog from an early age. Two other factors ought to be remembered. Firstly, certain breeds will have to allow the judge to examine inside their mouths – and this is not always an easy experience either from the point of view of the exhibit or of the judge. Secondly, the smaller breeds should be accustomed to standing on a table for the examination to take place. Once these problems have been overcome then your dog is ready to be entered for his first show. Only minor tricks are still to be learnt. These include keeping the dog alert during the entire time that he is present in the ring. Many people do this by the use of tit-bits in an attempt to hold the dog's attention in the ring, but this habit is not permitted in every country. Alertness all the time that the exhibit is present in the show ring is essential, since the judge's eye may wander to your exhibit at any time and impressions gained at a glance are so often long-lasting. Of course if your dog is not an eager showman, then it is as well to reserve his best performance for a moment when the judge is paying most attention to him. Experience is all that can perfect the technique of ringcraft, and the particular style of showing varies immensely from one breed to another. In some breeds 'topping and tailing' is permitted (topping and tailing means propping up the dog by holding his head and tail to make him show himself to best advantage), but in certain other breeds this habit is frowned upon. It is wise therefore to watch the techniques of fellow exhibitors quite closely before committing yourself to any course of action.

Finding a suitable show at which to make your first entry will not be difficult if you look in some of the canine journals which give fairly comprehensive lists of shows well in advance. Recognized shows fall into various categories. In the United Kingdom they can be divided into two main groups and thereafter into sub-divisions within these groups. Shows are either for one breed alone or are organized for all breeds by societies called General Canine Societies (as opposed to the Breed Clubs, who confine their activities to the one breed). Within these two main groups there are four further sub-divisions. Firstly there is the Sanction Show, which is in Britain the smallest type of show, being limited in the number of classes it may schedule to 25. No dog which has won more than a certain number of prizes can be entered. Then there is the Limited Show, which is slightly larger in number of classes and is, as its name suggests, limited to exhibitors who are members of the Canine Society organizing it. Membership of such a society can probably be obtained on payment of a small fee and without many other formalities. These Limited Shows are also barred to dogs which have already won several prizes, though this limitation is not so severe as in the case of Sanction events. Open Shows, as their name suggests, are for all comers, and you can expect to meet a considerable amount of strong competition at such events, though the classes themselves will probably be graded. Championship Shows, whose number in the United Kingdom is around thirty, excluding the one-breed shows, are really open shows at which can be won the most coveted award of all, the Challenge Certificate (or C.C. as it is commonly called). In order to become a Champion a dog must win three of these C.C.s under three different judges. Since there is only one Challenge Certificate on offer for each sex of each breed at a Championship Show, this is an honour which falls to exhibits only rarely – hence the great interest shown in these awards. Apart from the rarity of Challenge Certificates another factor which makes them so valuable is that the judge who is chosen to award these prizes has to be approved by the Kennel Club. In other countries methods of becoming a Champion vary, but it is still the case that the British Champion of many breeds is more sought after than those of any other country. In the United States dogs become Champions by winning a certain number of points; the number of points on offer at a particular show depends upon the number of dogs competing. But the difficulty of winning C.C.s in the United Kingdom is still so great as to make the award the most sought after in canine circles.

A Miniature Pinscher, Champion Bragan Go Gay, owned by Mrs M. H. Semple, poised on the table at a show waiting to be seen by the judge. Notice the fine slip lead which replaces the usual collar and lead for show purposes.

The question of which type of show to start the career either of a new exhibitor or of a new exhibit is often asked, and there are varying opinions on this topic. We feel on balance that the best type of show to attend first is a breed show. Firstly you will be able to meet fellow exhibitors of your own breed, to benefit from their advice and to see more specimens of your particular breed under one roof than would normally be possible at general canine club shows. Secondly, and perhaps this is more important, you will find that breed shows are inclined to be organized at a slightly more leisurely pace than variety events, and this will mean that the breeders and exhibitors will have more time to spare in giving helpful advice. Failing the availability of a breed show in your area, then the next best thing is the general championship show, which is likely to have classes specifically for your breed and will give you a chance to enter your exhibit with others of your breed. At the smaller all-breed shows the chances are that special classes confined to your breed, unless it is a popular one in the area, will not be available; even if they are it is quite possible that the judge may not be so experienced as he would be at a championship show. However availability will once again be the limiting factor in your choice, and you will possibly have to compromise by going to a local variety show and graduating from there to something further afield once you have gained more confidence and more experience. No two shows are alike and so, even if you do not enjoy your first event, do not be put off; next week's show will probably be quite different.

Having decided upon the show of your choice by referring to the canine journals you will then have to write for a schedule to the secretary of the event chosen. This will give you details of the classification. Do not be afraid to ask at your local training class for advice as to which classes to take part in. The information required on the entry form is very much the same as that appearing on the dog's registration papers and you must at this stage decide upon which classes you are going to enter. Definitions of classes, indicating which dogs are eligible for entry in which classes, will be contained in the schedule. In the United Kingdom classes such as Novice, Maiden and Graduate are graded according to the number of first prizes the exhibit has won. The Puppy and Junior classes are graded by age. Read the definitions carefully and if you are still having difficulty in deciding upon the classes in which to enter then do ring the secretary of the show, who will be only too pleased to help you. Entries will close anything up to five or even six weeks before the date of the show, and it is as well to enter fairly early so as to help spread the work load for the show management. Remember that once your entries have been submitted, you will receive no acknowledgement of their receipt until your exhibitor's admission ticket is sent to you, maybe several weeks later — almost at the time of the show day itself. In view of this fact many exhibitors at the larger shows send, along with their entries, a stamped and self-addressed post card which the show management can send back as acknowledgement of the receipt of the entries.

After a period of some months as a regular exhibitor at shows you will find that the need to write around to secretaries for show schedules will disappear as more and more clubs start to append your name to their mailing lists and as more and more schedules drop through your letter-box unrequested.

Grooming a dog for show purposes is very much a question of skill and experience, and as with other aspects of dog showing it varies from one breed to another. There are certain breeds which require little or no extra grooming before a show — these generally are the smooth-coated breeds. The rough-coated and some terrier breeds require a fair amount of grooming, and in some instances a certain amount of special trimming, if they are to be looked upon favourably by judges. Some of the long-coated breeds need a great deal of special care and attention, requiring bathing and grooming, as well as trimming and clipping, to put them in show condition. Generalizations are dangerous in this sphere, and so suffice it to say that the requirements of each breed should be studied carefully and that as much preparatory work as is possible should be carried out before the actual day of the show. Regular grooming and brushing should be enough to keep the dog in good overall condition, with perhaps a little extra attention during the weeks leading up to the show.

Trimming, as opposed to regular grooming, may be one of the reasons for the apparent shift in popularity in the show ring of some breeds as compared to others. It is found in the breeds that need trimming that it takes the newcomers longer to be able to present their exhibit really well.

For this reason, and considering the trend these days to expect fast results, some of the breeds which have become popular in recent years as far as show entries are concerned are the non-trimmed breeds. This is especially so in the terrier group, where Wire Fox Terriers have declined in entry numbers over the years, while Cairn Terriers and Border Terriers have increased in popularity in the show ring. This trend does not however seem to apply to the really long-coated dogs, where Afghan Hounds and Old English Sheepdogs are remarkably popular despite their obvious need for a considerable amount of care and attention.

You may wish to consider employing the services of a professional handler to take your dog to shows and to prepare him for exhibition. Generally speaking it is only the more wealthy exhibitors who use this type of service, though the costs cannot be defined adequately here. On the one hand the dog will be expertly handled and expertly presented and trimmed. On the other hand some of the thrill of winning with and showing your own dog will be missing if you use this service. There is no doubt that it does, however, make for a much more relaxed day for the owner of the dog who is visiting the show only to see his dog being shown and who has someone else to do all the work.

Another set of circumstances which varies from one breed to another is that which decides the amount and nature of impedimenta and equipment which need to be taken to the show – indeed this varies from one exhibitor to another, some being able to go to shows with no more than a comb in their pocket while others trudge into the show venue loaded down under vast quantities of paraphernalia. It is our opinion that in many breeds there is really no need for a great deal of work to be done on the day of the show itself, provided that adequate grooming has taken place in the time leading up to the event. If yours is a small breed which will require a good deal of attention at the show, then it may be helpful to take a small folding table for grooming purposes — something along the lines of a card table, though not of course with a card-table surface as this would gather hairs. Some breeds require chalking to clean their coats, and a supply of chalk will have to be taken to the show, though show managements are naturally inclined to discourage the use of this pervasive substance, especially in halls which have 'house proud' caretakers. The use of chalk to whiten a breed is not allowed under English Kennel Club Rules — it must, under the strictest observance of Rules, be removed from the dog's coat before he enters the ring. Judging by the disgraceful quantities of chalk which one sees at many shows however, it would seem that few people pay attention to this particular aspect of Kennel Club Rules. This laxness would seem to be a great pity, especially when it endangers the availability of venues for shows, when these are already scarce enough.

The Championship Shows and the largest of the Open Shows in Britain must all be benched shows at which dogs must be 'kennelled', or 'benched' as it is known, during the time when they are not actually being exhibited or exercised. This is done so that the show can present a spectacle for those members of the public who come along and pay for admission; it also serves exhibitors by giving them somewhere to leave their dogs while chatting to friends and seeing the rest of the show. If the show is a benched one, then you must remember to take with you a blanket, paper or some kind of covering for the bench on which your dog can lie and also a good strong collar and lead or preferably chain for leaving your dog on the bench without fear of his chewing himself free. Another finer show lead should be taken for showing the dog in the ring. Normally, these days, this will be made of leather, cord or nylon and will have attached some kind of adjustable slip collar.

Having tried to remember all the various factors that go to making the show run smoothly, having noted the time of judging and having calculated to arrive at the show around three-quarters of an hour before it is due to start, remember that your dog will be at the event to be shown – not only to the judge but also to the general public. Be sure that you have read the regulations of the show carefully and be especially sure that you do not have to be home by a time which would mean that you have to remove your dog from the show at a time before that permitted.

After your first visit to a show you will probably go home exhausted and perhaps disappointed, but we hope that you will have found, win or lose, that the experience has been worthwhile. It is a great opportunity to meet and speak to other exhibitors, to learn from their advice and to look to them for help and guidance in all respects, especially about particular points and peculiarities of your breed. Provided that you enter the show ring with the philosophy that it is a hobby in which you are taking part and not a business, and provided that you remember that judging is a subjective exercise and as such will vary from person to person and from week to week then you will be sure to enjoy your show going. You will be able to treat it merely as part of your main hobby of dog breeding, keeping it in perspective and at the same time learning more easily to take the rough with the smooth in a sportsmanlike way. This type of attitude will ensure that you will enjoy showing to the full and that you will contribute something in your turn to the dog fancy itself.

CHAPTER 8
Breeding Principles

The only good reason for breeding dogs is the desire to breed and raise good specimens of one's chosen breed; to do this the bitch's owner must acquire new knowledge and experiences and expend an appreciable amount of time and money and thought. All too often, however, people are reluctant to descend from the state of complete euphoria that their decision to breed a litter has placed them in, in order to acquaint themselves with such mundane matters as whether their particular bitch is fit to breed from, how to select a suitable mate, the bitch's needs during her pregnancy, the process of birth, and the puppies' and bitch's needs after it. As a consequence of this euphoria, therefore, unsuitable bitches may be mated to unsuitable dogs, the bitch exposed to unnecessary hardship and difficulty, and the ensuing puppies reared less well than they should be.

The primary requirement in the process is a bitch suitable to breed from. All conscientious breeders will select a potential brood bitch with very great care, for upon the quality of the puppies that she produces will depend their reputation and the extent of their future successes. The strength of any kennel lies in its bitches; only the best is good enough to breed from.

A brood bitch should be a typical specimen of her breed, free from any outstanding faults, mature but not aged, strong and in good condition. Her pedigree should contain good examples of the breed combined systematically in such a way as to produce good examples of their particular breed. If the bitch was originally obtained as a potential brood bitch she will have come from a kennel which, over a number of years, has produced good examples of the breed. If however the bitch was originally bought simply as a pet and possibly not even from a reputable breeder it is quite possible that she fails to meet these basic requirements, in

Mrs N. Kirby's Miniature Bull Terrier, Champion Kirbeon Bandmaster. Breeding smaller versions of a large breed has inherent dangers, but the attraction of small dogs is such that there is a determination amongst enthusiasts to overcome these difficulties.

which case we must stress that the bitch should not be bred from at all.

Stringent though the selection process for a bitch may be, a far higher standard of excellence should be set for the stud dog to which she is to be mated; nothing but the very best stud dog whose pedigree is compatible with the bitch's should be used. A novice breeder must beware of being beguiled by an owner's enthusiasm for his own dog and should base his judgment on the dog's success as a show dog or in some other activity as well as on his success as a sire. After all, what is the point of using a dog purely on the grounds of sentiment or of accepting a standard lower than the very best? Such an attitude is unlikely to better the chosen breed or to lead to any lasting satisfaction for the breeder.

Some owners may hesitate at paying what appears to them to be the high stud fee asked for the services of the best dogs, but a dog which does well in the show ring or in some other field is often the culmination of a long breeding programme or of a strict process of selection. Many puppies will have failed to achieve the standard of excellence demanded by his owner before a satisfactory dog was found. The dog's introduction to stud work will have been made with great care and thought and his entire existence will revolve round this activity. As a puppy he will have enjoyed the company of other dogs, for a solitary dog can often quite forget that he is a dog and become useless for stud work. When matured a proven steady bitch will have been made available to him, for a maiden or excitable bitch providing a difficult introduction to stud work could easily reduce or even destroy his enthusiasm for stud work. Only when the dog has sired two or three or even more litters of a quality which satisfies his owner will he be made available to a few selected bitches at public stud. For stud work a dog must be maintained in peak condition and be capable of and willing to mate quite recalcitrant bitches. His owner will be putting the dog's health and virility at some risk, for careful though he may be it is impossible to be as sure of the health and temperament of visiting bitches as he is of his own stock. Finally the owner is condemning himself to hours of exhausting work with unwilling bitches. For such a service a fee equal to half or three-quarters of the price of a good puppy is not at all unreasonable and certainly represents far better value for money than a much cheaper fee for the services of an inferior and unproven dog.

The selection of a particular dog as a mate for your bitch is a process which demands much thought and a detailed knowledge of the bitch's pedigree and those of the available dogs. Your aim should be to 'fix' as far as possible in the puppies the desirable characteristics of your breed, and to breed out any faults.

Puppies – like other animals and all forms of life – receive genes, the elements of heredity, from both parents. If both parents are genetically pure black (known technically as being homozygous for this characteristic), their puppies will all be pure black. But if one parent is pure black and the other pure white, the puppies will inherit genes for both colours, one of which will dominate the other (the weaker colour being called recessive). Let us suppose that black is the dominant, and that white is a fault. The puppies born from a mating of a pure black and pure white will all be black, but will carry the recessive white gene. These puppies are known as heterozygous blacks. If you mate two heterozygous blacks, you will get a surprise: a proportion of one in four of their puppies, given a sufficient number of litters, will be pure white, the same colour as one of their grandparents (or even a more remote ancestor), whose colour had apparently been lost. One in four of the grandchildren will be pure black (homozygous), the same as the other grandparent; and two in four will be heterozygous black, like their parents. Homozygous black and heterozygous black are identical in appearance; the difference is in the recessive genes, which are revealed only in their descendants. Other examples are given on page 94. With a knowledge of these basic principles you can predict the proportions of black and white puppies born from a mating between any two of these three different genetic types.

Not all characteristics have the simple dominant/recessive relationship of the example above. The dominance of one colour may be incomplete, in which case the puppies would be part-coloured; or a dilution factor may operate, resulting in grey puppies. Some genes are also linked with sex determination, or with other factors. Breeders are recommended to study specialist books on the subject.

Let us now consider how a knowledge of inheritance might be applied to a breeding programme to improve the quality of the puppies. If a new breeder acquires a bitch who, excellent though she herself may be, was the product

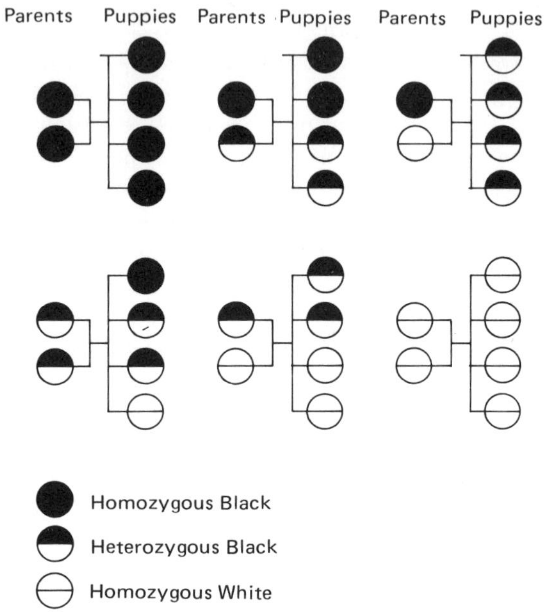

Diagram showing how colour and other characteristics are transmitted.

of an haphazard mating or whose pedigree contained dogs of which little was known, it would take three or even more generations before the owner would be able to predict what might be the result of any particular mating. If, however, the original bitch has been systematically bred, the owner should be able to predict what her puppies should be like and will be building on an established breeding programme with a firm foundation. For this reason, if for no other, it is wise to buy breeding stock from an established and successful strain. By this term we mean a group of dogs whose members are, to a greater or lesser extent, related to one another. The word 'strain' is much abused by dog breeders who are wont to talk glibly of their strain when no relevant relationship exists between the dogs in their kennel. A common prefix is no indication that a strain which has any reality in genetic terms exists. To produce and perfect a strain in dogs is the work of a lifetime and is achieved only by a dedicated few.

Basically there are two methods of breeding: from unrelated parents or from parents which, to a greater or lesser extent, are related. The former, from unrelated parents, cannot be regarded as a system. It is practised by human beings and tends to produce a divergence in the characteristics of the offspring rather than similarity. This method thus might produce offspring excelling in the required characteristics but it would depend on chance and since the probability cannot be assessed it is not used by serious livestock breeders.

In-breeding (or line-breeding as it is also called), from related stock, cannot introduce new elements and tends to intensify the characteristics, good or bad, of the parents in their offspring, who will tend to be very similar to one another. It is often supposed that such breeding encourages degeneracy and hereditary defects, but these faults are the result of uncontrolled consanguinity; livestock in-breeding is controlled. Serious breeders therefore welcome a knowledge of their stock's recessive factors and bad characteristics so that steps can be taken to build on

Champion Brackenfield Tertius, an Irish Setter owned by Miss S. J. Lennox. The Irish Setter is a good example of a breed which used to suffer from a hereditary disease, in this case night-blindness, that was virtually eliminated as a result of the combined efforts of breeders.

desirable traits and to eliminate undesirable ones.

In-breeding is a tool by which the purity of dogs for these good characteristics can be increased. Good, sound original stock and a fairly close degree of relationship between chosen mates are prerequisites of in-breeding. Parent to offspring, sister to brother, are the matings which fix type most quickly and which will tend to produce families of like individuals. The mating of half-brothers and half-sisters, grandparents and grandchildren or other such less close relationships may also be regarded as in-breeding; but such matings, although they will increase the purity of stock, will do so less quickly that closer matings. They will also obviate, or more likely postpone, the undesirable results of injudicious selection of breeding stock or faulty application of the system.

Occasionally out-breeding is necessary. For example, you may have a family which, though worth breeding from in all other respects, has too weak a muzzle in each individual. You must in this case find an unrelated dog whose muzzle is as near perfect as can be achieved and who is himself the descendant of good-muzzled dogs. The return to in-breeding between his descendants should fix this new point in your stock.

CHAPTER 9
Breeding and Weaning

A bitch comes into season, the time when she is ready to be mated, at fairly regular intervals. These intervals vary from breed to breed and from bitch to bitch, and only careful observation can really tell you when a bitch is in season. Most breeds have a fairly regular bi-annual cycle, but some (usually those with a fairly short history of domestication) only come into season annually. A period of illness, stress caused by a change of environment, the presence of quarrelsome kennel mates or a period of cold weather can all delay the onset of oestrus. In the same way warm weather or the companionship of other bitches in season will stimulate oestrus.

The first signs of an impending season are an increased restlessness on the part of the bitch, a tendency to urinate frequently, possibly in an attempt to advertise her condition to possible suitors, a swelling of the vaginal orifice and some resentment of the attentions of other dogs and bitches. As the heat progresses the swelling will increase and a bloodstained discharge, which varies in amount between individuals, will become evident. This discharge becomes less copious and less blood-stained over the next few days and it is during this stage that the bitch will be willing to accept the advances of a dog. Some bitches will stand for the dog over a period of several days but others may be ready only for a few hours. The bitch's behaviour towards dogs or even towards other bitches will advertise her willingness to be mated. She will flirt in the most brazen manner, will stand with her rear legs parted, her back arched and her tail curled to one side; an examination will show her vagina to be very swollen and soft. At this stage she must be taken to the dog as quickly as possible. If our advice has been taken and an outstandingly good stud dog is to be used, then you should have made prior approaches to the dog's owner to verify that the dog's services are available and to warn him that the bitch is due in season. The best dogs, particularly in some popuiar breeds and at certain times of the year, cannot accept every bitch, or may be unavailable for some other reason.

On the appointed day the bitch should be taken to the dog, seldom the other way round. Most dogs are creatures of habit and familiar surroundings are helpful to many stud dogs. Neither the bitch nor the dog should have been fed for a short time prior to the mating, for a full stomach can make the process unnecessarily uncomfortable and, sometimes, impossible. The owner of the bitch should follow the instructions of the stud dog's owner precisely, for many dogs develop very strange idiosyncrasies for which their owner will be prepared and make allowance. After the initial introduction, which gives the dog and bitch an opportunity to assess one another, has taken place, the bitch's owner may be asked to restrain her head and at all costs to prevent her from sitting when the mating is taking place. A dog's penis is erected by means of a small bone, the *os penis,* which can be broken by a bitch sitting when mating is taking place. If this small bone is broken the dog becomes totally useless for stud work, since he will be unable to mate a bitch successfully. When the dog mounts the bitch he will clasp her firmly round the abdomen with his front legs and will remain in this position until penetration has taken place, which for a small or maiden bitch may be accompanied by a sharp pain. Once penetration has taken place the dog will begin to turn himself, a process an experienced dog can achieve with only token assistance from his handler providing that the bitch is held steady. The dog will lift one foreleg over the bitch and then move round until he can lift a hind-leg over the bitch's quarters (if the bitch is big or unsteady he may need some gentle help to achieve this), after which the tie is complete. The two dogs will be tied back to back and remain so for anything up to an hour. During this time the dog is locked into the bitch by a large swelling at the tip of his penis, and any sudden movement on his part or the part of the bitch could injure both quite severely, so that the pair should be quietly restrained during the entire process. Mainly for the comfort of their owners toy dogs are very often mated on tables, but with larger dogs a couple of camp stools may help to alleviate some of the discomfort that their handlers may otherwise feel.

If yours is a maiden bitch or one which has shown a reluctance to previous matings, if the dog is used only infrequently, or if the mating was unsatisfactory in some way, then two matings may be advisable. The second should never take place more than forty-eight hours after the first, as the puppies conceived at the second mating may be born prematurely or those conceived at the first may be injured by a delayed birth. However for a normally healthy vigorous pair mated at the right time there seems to be little that can be gained by a second mating.

When the mating has taken place the dog should be taken back to his kennel and a check made soon after that his penis has withdrawn into its sheath. He should be washed with mild disinfectant, given a drink and allowed to rest quietly for a time. The bitch too should be taken gently away and during the next few days should be kept quiet and calm and any stress or excitement avoided. This time is critical to conception and to the size of the ensuing litter.

During the early weeks of her pregnancy a bitch needs no special attention; her food intake need not be increased nor her exercise decreased. Indeed it is of the utmost importance that, during this period, she should be kept in the peak of health and on no account allowed to get fat. Obesity reduces the chances of conception taking place, puts additional strains on a bitch's constitution during her pregnancy and creates unnecessary dangers and difficulties during the process of birth. Exercise can be maintained at a normal level until pregnancy is well advanced, so that the bitch is in the best possible condition to discharge her allotted duties.

During the later stages of her pregnancy violent exercise

should be avoided, but steady exercise, perhaps on a lead, is best continued until whelping is imminent. Her diet should, during these later stages, consist of highly nutritious foods, which may be given in smaller more frequent meals than she has been accustomed to so that her stomach is never overloaded. Good-quality meat, milk, fish, cooked eggs, perhaps with the addition of some cod liver oil, calcium or preferably a balanced calcium-phosphorus supplement will all help to ensure that the bitch is well prepared for the task that lies ahead of her. She should, as far as is possible, be protected from the risk of disease or injury at this time. She should certainly not come into contact with any sick dogs and it is advisable to keep her away from dog shows or any dogs which might have been in contact with disease. If the bitch is of a breed whose normal work involves some risk of injury, she should be kept away from work from the time of mating until she has recovered from her litter.

A suitable place in which to whelp should be prepared well before the puppies are due so that the bitch can become accustomed to it. If the bitch is of a toy breed or is to whelp in a cold climate or during winter she and her puppies will need some supplementary heat, for, particularly during the first few days of a puppy's life, heat is as or more important than food if he is to survive. The bed which is provided for the bitch should be big enough for her to lie down at full stretch, should be made of a material which is warm but easily cleaned and should have sides of sufficient height to protect the pups from draughts and to prevent them from falling out of bed. For larger breeds or those which are inclined to be clumsy and are, therefore, likely to injure the puppies by lying on them a creep rail should be provided around the box. It is advantageous if the box has one side which can be removed or lowered to allow the growing puppies to explore outside the bed.

The bedding is dependent on the preference of the owner, the breed and availability. It should be disposable or easily cleaned and not envelop or entwine the puppies and so prevent them from getting to their dam. Many breeders find old clean newspapers are light and warm and make an ideal bedding for their dogs at this time. They are easily disposed of and equally easily replaced. Furthermore they seem to satisfy a bitch's urge, particularly when birth is imminent, to arrange her own bed. A bitch will often tear her bedding into small shreds and make a nest to her own liking. This is perhaps one of the first signs that she is getting ready to have her litter. It is possible to buy shredded paper quite cheaply and, providing that a source which can guarantee the cleanliness of its product can be found, this makes an excellent bedding at all times. Straw or woodwool are often used, but inside the home they are indescribably pervasive, while the cleanliness of straw cannot be guaranteed. Blankets, sheets and towels are good for older puppies but can accidentally cover a small or very young puppy with disastrous results and, of course, have to be kept scrupulously clean rather than simply replaced.

A bitch is usually reluctant to whelp in unaccustomed surroundings and so her bed should be prepared and placed where she is expected to whelp well before birth is imminent. From time to time one meets a contrary bitch who will insist on following her own inclinations regarding a suitable place to give birth, invariably choosing a place which is neither suitable nor convenient to her or to her owner, but she can usually be persuaded to accept an alternative which has been provided and to which she is already accustomed. Its temperature should be raised to a steady 70–75 °F (21–24 °C) shortly before whelping is due.

Even the most experienced dog breeders are as anxious as the novice to detect some signs that the bitch is in whelp. Some breeds are easier than others to get into whelp; while within a breed some strains may be recognized as being more fertile than others. We regret that it is not customary to publish a dog's fertility record in the same way that those of stallions are published; in this way it would be possible to assess the possibility of achieving a litter by a certain dog and to equate this with the performance of other possible mates. In the absence of this information one can only rely on any indications which may be shown by the bitch herself. During her season the bitch's teats and vulva will be noticeably swollen and will often remain so if she is in whelp. Sometimes her appetite will become somewhat capricious or she may be less inclined to play boisterously than hitherto. About three weeks after conception it is sometimes possible for a vet or other experienced person to feel the developing embryos. During the last week of her pregnancy the movement of the unborn puppies can often be detected by gently pressing the fingers against the bitch's abdomen. By this time the bitch will be considerably enlarged but this in itself is no proof that she is pregnant, for it is possible for her to have a false or phantom pregnancy which exhibits most of the symptoms of a real one but, obviously, none of those exhibited by tactile examination. Such 'pregnancies' may disappear before the supposed whelping date or the bitch may even be so deluded as to make a bed and 'give birth' to her imaginary litter.

Some bitches, not just frustrated mothers but even experienced brood bitches, may have a phantom pregnancy after each season, the only danger being of abscesses following an unrelieved secretion of milk. A bland diet of fish, vegetables, tripe and so on, a reduction in her intake of liquid and an increase in hard exercise will often persuade a bitch to forego her imaginary litter and return to normal.

In an age of obsessive cleanliness there is a general abhorrence of internal parasites, particularly worms. It is true that a large infestation can weaken a bitch or stop a puppy from thriving but it is also true that an unsuitable or unwisely used worming agent can do just as much harm. If there is good reason to believe that the bitch has worms then she should be treated before she is mated, never afterwards, except when so advised by a vet.

The bitch will be due to have her puppies sixty-two or sixty-three days after she has been mated. For the birth, sterilized scissors may be needed, warm towels, plenty of clean bedding, warm, mildly antiseptic water, the vet's telephone number and perhaps above all the knowledge that everything that can be done for the bitch has been done. Self-recrimination after a bitch has lost all her puppies or has herself died as a result of some lack of care or thought on the part of her owner can, and should, be very bitter indeed. Finally, supervision of the entire process of birth is a necessity. A healthy bitch will normally whelp naturally and easily but things can go wrong and a watchful eye can save the bitch much unnecessary suffering.

Should whelping be delayed more than two or three days after the normal sixty-two or sixty-three, veterinary advice should be sought. The normal signs that the birth is approaching are fairly obvious. The bitch may become reluctant to take any exercise a couple of days before the birth is due, may make a nest by tearing up and arranging the bedding in her chosen place, and may become increasingly restless as the birth approaches. When labour is imminent her temperature will drop by a couple of degrees from the normal 101·5°F (38·6°C), and her vagina will become soft and enlarged. The bitch may stay in this condition for a day or two, without any need for alarm on the part of the owner, providing she is showing no signs of discomfort or distress, before progressing to the first stage of labour.

The first stage of labour is characterized by irregular contractions, which may not always be obvious except to an experienced observer. During this stage the bitch will be in some discomfort and, as a consequence, somewhat restless. The genital passages become enlarged and lubricated and not until this process is complete can the puppies be born. This stage may only last for a few hours, but, providing that the contractions have not become rhythmic or powerful, and that the bitch is not in acute discomfort or exhausting herself, no undue alarm need be felt if it extends to twenty-four hours.

As soon as the genital passages are sufficiently softened,

dilated and lubricated to allow the birth to procede, labour can move into the second stage. The contractions will become more powerful, rhythmical and obvious to the eye. These powerful muscular spasms gradually force the lowest puppy through the genital passages until the bag in which the puppy is enclosed appears at the vulva, probably to disappear again as the contraction passes. Gradually the puppy will be forced further out until bag, puppy and the attached placenta are free. At this stage the bag may already have broken or been torn by the bitch with her teeth at the moment of birth, allowing the puppy to take its first breath, but if the bag has not broken and the bitch is showing no interest as a result of being tired or a little bewildered, then it should be torn by you, and gently removed and the puppy rubbed vigorously with a warm towel until it is moving and breathing freely. The bitch will normally bite through the umbilical cord which joins the puppy to the placenta, although some maiden or nervous bitches may be reluctant to do so and others of the ultra short-faced, brachycephalic breeds may have some difficulty in severing it. Again the bitch should be helped by your squeezing the contents of the cord towards the puppy with the fingers, thus allowing as much blood as possible to go into the puppy, and then cutting the cord about 1½ inches away from the puppy with sharp sterilized scissors.

After the bitch has severed the cord, removed the bag from the puppy and given it a vigorous licking, she will normally procede to eat the placenta. Distasteful though this action may be to some people, she should be allowed to do so because it contains a substance which stimulates the production of milk.

After the first puppy has been born the bitch may rest quietly for a while or may procede immediately to give birth to the second puppy. The litter is usually all born within a space of five to ten hours, with perhaps, as the bitch tires, a noticeably longer interval between the later puppies. During her exertions the bitch may benefit from a drink of warm milk to which glucose has been added, but apart from this she need be given nothing other than constant attention and reliable assistance, if needed.

Novice breeders need not be alarmed at a considerable bloodstained discharge after the puppies are born. A slight discharge is quite normal for ten days to a fortnight. After whelping the temperature is often abnormally high by a couple of degrees, but within forty-eight hours it should have returned to normal. The bitch's motions will be black and glutinous for a day or two after the birth and at this time she will be susceptible to diarrhoea and to an unwanted rise in temperature if she is given too much rich food too soon after the birth of her puppies. For the first day after whelping she should be given liquid food only, to give the digestive organs time to get back to normal.

Most bitches will settle with their puppies and quickly become adept at nosing them about the bed, cleaning and feeding them and it is often difficult in the first few days to persuade a bitch to leave her charges in order to eat or relieve herself.

Not all births however will be so uneventful. Some breeds, notably those which fashion has decreed should have abnormally large heads and abnormally slender rear ends, invariably have more difficulty in giving birth and often need surgical assistance. With more normally developed breeds trouble is less frequently encountered, but when it does arise it is essential that you should be able to recognize the signs and to act quickly to give or summon whatever aid is required.

The two primary causes of whelping difficulties are inertia on the part of the bitch, and some mechanical reason resulting from an obstruction within the bitch or from a malpresented puppy. Inertia of the uterus means an absence of the facility to contract the uterine muscles rhythmically in order to expel the puppy. This can happen if an elderly bitch or one not in good physical condition is bred from or can result from poor feeding or inadequate exercise. Uterine inertia can also result from exhaustion after an excessively prolonged labour.

A pelvis smaller than the puppy's head will make a normal birth impossible, and an abnormally large puppy or one lying in an abnormal position can make the birth very difficult. If the puppy is lying transversely or the head is bent on the chest the puppy will have some difficulty in passing through the pelvis. Such abnormal presentations are usually beyond the skill of a novice to deal with and skilled assistance should be sought as quickly as possible. Some help however can be given even by the inexperienced person. If a probing finger can feel the puppy and ascertain how it is lying then it will often be possible either to ease it out by pulling gently (but only during a contraction) or by changing its position slightly so that the obstruction to its birth is removed. The most frequently encountered difficulty is the breech birth, in which the puppy is presented tail first rather than head first. By gently easing and turning the puppy on its axis it can often be freed.

For a novice breeder the birth of the first litter may be a traumatic experience; but this is as nothing compared to the stress to which both bitch and puppies have been subjected. A healthy bitch after an hour or two's sleep and a warm drink will quickly overcome this stress and providing that she is well cared for during the ensuing weeks need suffer no after-effects. The puppies must be given every assistance to overcome the shock of the birth itself, to make the adjustment of fending for themselves and to withstand the tremendous drop in temperature which they will have encountered. There is a lot of evidence that the primary cause of very many puppies failing to survive the first few days after birth is sheer lack of warmth. Such a small creature loses body heat very quickly and will complain plaintively for a while, until as his temperature drops this response is dulled and the owner may assume that all is well. Even at this stage access to the mother's warm milk can quickly revive him, but beyond this stage he very quickly loses the ability to suck and then he can only be saved by being placed in a higher temperature. We cannot stress too strongly that for the first few days of life warmth, a constant and minimum temperature of 75 °F (24 °C) is as great, or even greater, in importance as food.

There are various ways of providing satisfactory heat for puppies. It is possible to raise the temperature of the whole room or kennel, in which the puppies are to live; but a far more usual and economical method (though not one without some disadvantages) is to provide localized heat from an infra-red or dull emitter source such as is commonly used for rearing other small livestock. The disadvantages of these heaters are that they can dehydrate the puppies and that since the heat is very localized the puppies can get outside its influence and so quickly become chilled without the owner noticing. The final choice will depend on individual preference and circumstances.

A diet with a high protein content is fundamental to a well-reared litter, but during the first few days after whelping, before the puppies are making great demands on their dam for food, the bitch should be kept on a bland diet of fish and milky foods. Those sold for babies are excellent and, surprisingly, often cheaper than those prepared specifically for dogs.

After two or three days the puppies will be becoming more vigorous and demanding; the bitch should be given raw meat, which is rich in protein and stimulates the production of milk. Milky foods which will help to replace the drain on the mother's reserves of calcium should be continued. If four meals a day are given, two of meat and two of milk-based foods, and the bitch has constant access to clean drinking water she should be well fortified. Whether or not it is necessary to supplement such a diet is a subject which breeders tend to discuss endlessly. Many breeders rely on a completely natural diet, while others seem to have great faith in these proprietary supplements, without there being any noticeable difference between the puppies produced by one system or the other. So once again it is up to the individual to follow the line which gives the best results for his particular breed.

About ten to fourteen days after their birth the puppies eyes will begin to open and after this stage they will rapidly become more independent and adventuresome. During their third week of life they can be taught to lap at one of the proprietary milk foods available, and to eat a little finely scraped raw meat, so that by the end of that week they should be eating four small meals a day as well as suckling from their dam. After about a month the bitch's enthusiasm for her litter will probably begin to wane and her diet can be reduced accordingly, so that by the time the puppies are six weeks old they are independent of their dam and their food has been increased in proportion to their rate of growth.

The bitch will, or should be encouraged to, leave her puppies for short periods almost from a few days old. These periods should be gradually lengthened so that by the time the pups are five weeks old she is visiting them only three or four times a day to feed them and sleep with them at night. This gradual weaning process should be continued, with the bitch seeing less and less of her pups, until at seven weeks old they should be completely independent of their dam and sleeping apart from her. Even before this stage it is a good idea to provide the bitch with a separate bed, inaccessible to the puppies, so that she can wean them herself. If the bitch is given every opportunity to make this a gradual process she will do it easily and naturally and the puppies will be slowly and gently accustomed to leading an independent life.

After the sixth week the puppies may be given four or five meals a day, with as much variety being introduced into their diet as possible, so that when they go to their new homes at eight or nine weeks of age they are less likely to be upset by the change. At this stage the puppies appetites will be quite prodigious, and the cost of properly feeding a litter, particularly that of a large breed, can be considerable. The diet should contain fortified milk foods, beaten egg yolk in milk, minced meat and fish, and perhaps some finely ground meal or toasted brown bread which has been thoroughly soaked in gravy, but the prime food is and should remain good-quality raw meat. At this stage in its life an active, growing puppy might easily consume an ounce of food for each pound of body weight – roughly twice the amount required by an adult dog – and he should not be stinted. Exercise need not be a problem at this age, as a growing puppy will get all that he requires by wrestling and playing with his litter mates. He will tire easily and sleep a lot and he should be allowed to do so.

Some breeds are notoriously prolific, often producing ten or even a dozen puppies in a litter, but none should be expected to rear such a litter. Among a litter of this size there will inevitably be some which at birth are noticeably inferior to the others – perhaps they are mismarked or weaker than the others in the litter – and these should be put down at birth. It does no good to a breed or to the reputation of a breeder to rear inferior puppies, and none should make soft heart or the possibility of profit an excuse for not destroying them at birth and leaving the bitch with a litter of six or seven puppies which she can rear comfortably and well.

In exceptional circumstances, perhaps because of the death of the bitch or her refusal to accept any responsibility for her puppies, it may be necessary to rear a litter by other than normal means. Two ways are possible: the easiest and by far the most satisfactory is to acquire a foster mother for the puppies. A bitch with a very small litter might be pressed into service or one with milk after a false pregnancy might be prepared to help, but the breeder who is able to locate a suitable bitch in time to be of use is very fortunate. The alternative, hand-rearing a litter, involves maintaining them at a constant temperature not less than 75 °F (24 °C), feeding them at two-hourly intervals through the day and night, encouraging them to pass water by simulating the bitch's licking of the abdomen, by rubbing gently with a small piece of cotton-wool, and thoroughly cleansing them afterwards, with a small piece of warm damp cotton-wool. The puppies should be fed a proprietary bitch's milk substitute, which can be easily obtained at good pet stores or large chemists, using the sort of small feeding bottles which are used for premature babies. After the feed the puppies and their bed must be thoroughly cleansed and the puppies allowed to rest until their next feed.

The nails of a well-fed litter grow surprisingly quickly and can cause acute discomfort to the bitch as the puppies knead her to persuade her to release milk. Nails should always be kept well trimmed. At about four days old the puppies of certain breeds should have their dew claws removed. After the bitch has been taken away, the dew claws and the rudimentary thumb on which they are set can be cut off with a pair of sharp sterile surgical scissors. The puppies may cry out but will quickly return to feeding when they are returned to their dam. Bleeding can be reduced by the application of Friars Balsam or a weak solution of permanganate of potash. Some breeds should also have their tails docked. This operation must take place during the first week of the puppy's life but calls for more skill, if only to ascertain just how much tail to leave on each puppy. It is wise to get experienced help for the first attempt.

Finally, it is best to get your puppies accustomed to being handled gently. Impressions gained at this early age can make a big difference to their subsequent development and to their attitude to human beings.

APPENDIX 1
The World's Kennel Clubs

AUSTRALIA
Australian National Kennel Council,
Royal Show Grounds, Ascot Vale, Victoria, Australia.

BARBADOS
Barbados Kennel Club,
Avondale, Bank Hall, St Michael, Barbados, W. I.

BELGIUM
Société Royal Saint-Hubert,
Avenue de l'Armée 25, B-1040, Brussels, Belgium.

BERMUDA
The Bermuda Kennel Club, Inc., P.O. Box 1455,
Hamilton, Bermuda.

BRAZIL
Brazil Kennel Club,
Caixa postal, 1468, Rio de Janeiro, Brazil.

BURMA
Burma Kennel Club,
Room No. 10, 342 Maha Bandoola Street, Rangoon, Burma.

CANADA
Canadian Kennel Club,
111 Eglinton Avenue East, Toronto 12, Ontario, Canada.

CARIBBEAN
The Caribbean Kennel Club,
P.O. Box 737, Port of Spain, Trinidad.

CHILE
Kennel Club of Chile, Casilla 1704, Valparaiso, Chile.

COLOMBIA
Club Canino Colombiano,
Carrera 7a, No. 84-61, Apto 101, Bogota, D.E., Colombia.

DENMARK
Dansk Kennelklub,
Norrebrogade 40, 2200 Copenhagen, Denmark.

EAST AFRICA

East African Kennel Club,
P.O. Box 511, Nairobi, Kenya.

EIRE

Irish Kennel Club, 4 Harcourt Street, Dublin 2, Eire.

ENGLAND

The Kennel Club,
1, Clarges Street, London, WIY 8AB, England.

FINLAND

Suomen Kennelliitto-Finska Kennelklubben,
Bulevardi 14A, Helsinki, Finland.

FRANCE

Société Centrale Canine,
3 Rue de Choiseul, Paris 2, France.

GERMANY

Verband für das Deutsche Hundewesen (VDH),
Schwanenstrasse 30, Dortmund, W. Germany.

GUERNSEY

Guernsey Dog Club,
Myrtle Grove, St Jacques, Guernsey, C.I.

HOLLAND

Raad van Beheer op Kynologisch Gebied in Nederland
Emmalaan 16, Amsterdam, Holland.

HONG KONG

Hong Kong Kennel Club,
3rd Floor, 28B Stanley Street, Hong Kong.

INDIA

Kennel Club of India,
Kenhope, Coonoor 1, Nilgris, S. India.

JERSEY

Jersey Dog Club,
La Huppe, Birches Avenue, St Saviour, Jersey, C.I.

MALAYA AND SINGAPORE

Malaysian Kennel Association,
P.O. Box No. 559, Kuala Lumpur, Malaya.

MALTA G.C.

Malta Kennel Club,
12 Our Saviour Street, Sliema, Malta G.C.

MONACO
Société Canine de Monaco,
Palais des Congrès, Avenue d'Ostende, Monte Carlo, Monaco.
NEW ZEALAND
New Zealand Kennel Club,
P.O. Box 523, Wellington, CI, New Zealand.
NORWAY
Norsk Kennelklub,
Bjorn Farmannsgate 16, Oslo, Norway.
PAKISTAN
The Kennel Club of Pakistan,
Fortress Stadium, Lahore, West Pakistan.
PORTUGAL
Cluba Portuguese de Canicultura,
Praca D Joao da Camara 4-3, Lisbon 2, Portugal.
SCOTLAND
The Scottish Kennel Club,
18 Ainslie Place, Edinburgh, EH3 6AX, Scotland.
SOUTH AFRICA
Kennel Union of Southern Africa,
P.O. Box 562, Colonial Mutual Buildings, 106 Adderley Street, Cape Town, South Africa.
SPAIN
Real Sociedad Central de Fomento de las Razas en Espana,
Los Madrazo 20, Madrid, Spain.
SWEDEN
Sveska Kennelklubben,
Box 1121, 111 81 Stockholm, Sweden.
SWITZERLAND
Schweizerische Kynologische Gesellschaft,
Postfach 2307 3001 Berne 1, Facher, Switzerland.
URUGUAY
Kennel Club Uruguayo,
Avda, Uruguay 864, Montevideo, Uruguay.
U.S.A.
American Kennel Club,
51 Madison Avenue, New York, U.S.A.

APPENDIX 2
Whelping Calculator

Date of whelping is given immediately below the date of service

Jan.	1	2	3	4	5	6	7	8	9	10	11	12	13	14	15
Mar.	4	5	6	7	8	9	10	11	12	13	14	15	16	17	18

Jan.	16	17	18	19	20	21	22	23	24	25	26	27	28	29	30	31
Mar.	19	20	21	22	23	24	25	26	27	28	29	30	31	1	2	3 Apr.

Feb.	1	2	3	4	5	6	7	8	9	10	11	12	13	14	15
Apr.	4	5	6	7	8	9	10	11	12	13	14	15	16	17	18

Feb.	16	17	18	19	20	21	22	23	24	25	26	27	28
Apr.	19	20	21	22	23	24	25	26	27	28	29	30	1 May

Mar.	1	2	3	4	5	6	7	8	9	10	11	12	13	14	15
May	2	3	4	5	6	7	8	9	10	11	12	13	14	15	16

Mar.	16	17	18	19	20	21	22	23	24	25	26	27	28	29	30	31
May	17	18	19	20	21	22	23	24	25	26	27	28	29	30	31	1 Jun.

Apr.	1	2	3	4	5	6	7	8	9	10	11	12	13	14	15
Jun.	2	3	4	5	6	7	8	9	10	11	12	13	14	15	16

Apr.	16	17	18	19	20	21	22	23	24	25	26	27	28	29	30
Jun.	17	18	19	20	21	22	23	24	25	26	27	28	29	30	1 Jul.

May	1	2	3	4	5	6	7	8	9	10	11	12	13	14	15
Jul.	2	3	4	5	6	7	8	9	10	11	12	13	14	15	16

May	16	17	18	19	20	21	22	23	24	25	26	27	28	29	30	31
Jul.	17	18	19	20	21	22	23	24	25	26	27	28	29	30	31	1 Aug.

Jun.	1	2	3	4	5	6	7	8	9	10	11	12	13	14	15
Aug.	2	3	4	5	6	7	8	9	10	11	12	13	14	15	16

Jun.	16	17	18	19	20	21	22	23	24	25	26	27	28	29	30
Aug.	17	18	19	20	21	22	23	24	25	26	27	28	29	30	31

Jul.	1	2	3	4	5	6	7	8	9	10	11	12	13	14	15
Sep.	1	2	3	4	5	6	7	8	9	10	11	12	13	14	15

Jul.	16	17	18	19	20	21	22	23	24	25	26	27	28	29	30	31
Sep.	16	17	18	19	20	21	22	23	24	25	26	27	28	29	30	1 Oct.

Aug.	1	2	3	4	5	6	7	8	9	10	11	12	13	14	15
Oct.	2	3	4	5	6	7	8	9	10	11	12	13	14	15	16

Aug.	16	17	18	19	20	21	22	23	24	25	26	27	28	29	30	31
Oct.	17	18	19	20	21	22	23	24	25	26	27	28	29	30	31	1 Nov.

Sep.	1	2	3	4	5	6	7	8	9	10	11	12	13	14	15
Nov.	2	3	4	5	6	7	8	9	10	11	12	13	14	15	16

Sept.	16	17	18	19	20	21	22	23	24	25	26	27	28	29	30
Nov.	17	18	19	20	21	22	23	24	25	26	27	28	29	30	1 Dec.

Oct.	1	2	3	4	5	6	7	8	9	10	11	12	13	14	15
Dec.	2	3	4	5	6	7	8	9	10	11	12	13	14	15	16

Oct.	16	17	18	19	20	21	22	23	24	25	26	27	28	29	30	31
Dec.	17	18	19	20	21	22	23	24	25	26	27	28	29	30	31	1 Jan.

Nov.	1	2	3	4	5	6	7	8	9	10	11	12	13	14	15
Jan.	2	3	4	5	6	7	8	9	10	11	12	13	14	15	16

Nov.	16	17	18	19	20	21	22	23	24	25	26	27	28	29	30
Jan.	17	18	19	20	21	22	23	24	25	26	27	28	29	30	31

Dec.	1	2	3	4	5	6	7	8	9	10	11	12	13	14	15
Feb.	1	2	3	4	5	6	7	8	9	10	11	12	13	14	15

Dec.	16	17	18	19	20	21	22	23	24	25	26	27	28	29	30	31
Feb.	16	17	18	19	20	21	22	23	24	25	26	27	28	1	2	3 Mar.

INDEX